Energy Security

Nikolai Mouraviev · Anastasia Koulouri
Editors

Energy Security

Policy Challenges and Solutions for Resource Efficiency

palgrave
macmillan

Editors
Nikolai Mouraviev
Dundee Business School
Abertay University
Dundee, UK

Anastasia Koulouri
Dundee Business School
Abertay University
Dundee, UK

ISBN 978-3-030-01032-4 ISBN 978-3-030-01033-1 (eBook)
https://doi.org/10.1007/978-3-030-01033-1

Library of Congress Control Number: 2018955598

Cover credit: W. Michael Wiggins/Alamy Stock Photo

This Palgrave Macmillan imprint is published by the registered company Springer Nature Switzerland AG
The registered company address is: Gewerbestrasse 11, 6330 Cham, Switzerland

To my father, for always being there.
Anastasia Koulouri

Contents

Notes on Contributors

Dr. Alison Ashby is a Lecturer in Operations Management at Plymouth Business School, UK where she researches and teaches sustainability in the context of operations and supply chains. Her strong interest in sustainability evolves from direct experience of working within the UK textile industry and identifying mechanisms to address both environmental and social issues across the global supply chain. She has published in major operations, supply chain and sustainability journals and books, and her current research interests include closed loop supply chains and the circular economy, renewable energy and resource efficiency and reshoring for sustainability.

Dr. Zhibek Bimagambetova is an Associate Professor and the Director of Centre of Excellence at Al-Farabi Kazakh National University. Her research focuses on the social and psychological aspects of the new energy infrastructure. Her recent research activities include a project on impact of global climate change on land degradation in Kazakhstan (in cooperation with the British Council) and a project on promoting clean energy in resource-rich countries (in cooperation with the University of Reading and Kazakh Economic Research Institute).

Dr. Marat Karatayev is a Newton Al-Farabi Energy Research Fellow at the Energy Technology Research Institute, University of Nottingham, UK. He is also a Research Associate with the Kazakh Institute for Economic Research and UNESCO Department on Sustainable Development at Al-Farabi Kazakh National University. With Ph.D. in Urban, Energy and Environmental Planning, his research interests include clean fuel and energy technologies, sustainable cities, water-food-energy nexus and sustainability indicators. He has participated in a range of energy projects sponsored by the European Commission, the British Council, and the Science Fund of Kazakhstan.

Inna Kostrytska is a Senior Associate, Advocate with Jeantet, an independent law firm based in France and operating worldwide. She graduated from Yaroslav Mudryi National Law University, one of leading law schools in Ukraine. She also earned a Master degree in Finance. Inna specialises in energy law, M&A, and is in charge of the energy law practice of Jeantet's Kyiv office. Inna is an author of numerous articles and actively participates in the development of Ukrainian law as a member of working groups with the European Business Association, the American Chamber of Commerce in Ukraine, other business and professional associations.

Dr. Anastasia Koulouri graduated from the University of Strathclyde, UK with a Ph.D. in Management Science. Previously she has held posts at the University of Strathclyde, the Technical University of Helsinki and the Belgian Nuclear Energy Research Centre. Currently she is a Lecturer at the Dundee Business School, Abertay University, UK. Her research interests include energy security in resource-rich countries and sustainable development of transitional economies.

Dr. Nikolai Mouraviev is a Senior Lecturer at the Dundee Business School, Abertay University, Scotland, UK. Previously he has held teaching positions at KIMEP University, Kazakhstan, Wayne State University, US, and Viterbo University, US. A large part of his research focuses on public–private collaboration in transitional nations and he has published extensively in this area including a monograph (with Palgrave Macmillan in 2017) and an edited volume (with Cambridge

Scholars Publishing in 2017). In addition, his research embraces energy policy and governance.

Dr. Galiya Movkebayeva is a Professor of International Relations and Economics, and the Director of Centre for Eurasian Studies at Al-Farabi Kazakh National University. Her research is focused on the energy security and geopolitics, energy and democracy in former Soviet countries, climate and energy policies of the European Union. Her research work has been supported by grants from the European Commission, the German Science Foundation, Konrad Adenauer Foundation and European Academy of Diplomacy. Currently she is a coordinator of a project on energy security and energy policymaking framework in the Eurasian Economic Union, funded by Kazakhstan's Ministry of Education and Science.

Mihaela Grubišić Šeba, Ph.D., CFA is a Research Associate and a senior consultant with the Institute of Economics, Zagreb, Croatia with 15 years of experience in research/consulting and in the banking sector. She has published more than 40 academic papers and completed 30 studies in the fields of project, corporate and public finance. Her professional interest is in analysing financial options available to enterprises and in financial appraisals of projects in transport infrastructure, energy efficiency, and restructuring of private and public enterprises.

Carl Tidemann is a Ph.D. Scholar from the Fenner School of Environment and Society, Australian National University. His research focuses on the decarbonisation of the Australian electricity sector and the role of different actors in that task. He has an interest in all facets of policy making and has published on the integration of environmental concerns in international trade. He received his B.Sc. (Hons. 1st class) from the Australian National University. As well as research, he provides policy advice to a number of non-government organisations working to promote ambitious climate and energy policy in Australia.

Dr. Umut Turksen is a Professor in Law at Coventry Law School, UK. He has extensive expertise in the EU energy security and financial crimes law. He has provided consultancy to prestigious international businesses and government projects. These include technical assistance programmes to multinational corporations, ministries and international organisations. Professor Turksen's current research focuses on two themes: solidarity in the EU external energy trade and on countering financial crimes. He is currently leading a EU H2020 project, PROTAX, that aims to provide solutions for prevention and prosecution of serious tax crimes in the EU.

List of Figures

List of Tables

1

Introduction: Towards a Novel Conceptualisation of Energy Security

Nikolai Mouraviev and Anastasia Koulouri

Energy continues to draw considerable attention from citizens and governments across the globe in both industrialised and developing countries. Many nations are concerned with the continuity of energy supply to ensure that the needs of their economies, businesses and households are fully served. To this end, governments form their national energy policies, and energy security often becomes an integral part of these policies and remains an ongoing concern. What exactly is the nature of this concern? How do governments address energy security concerns? These questions have formed a foundation for this book and also induced the authors to look at the critical factors that underpin government energy policy and actions.

When governments talk about energy security, are governments concerned about the right thing? The reason behind this question is that,

N. Mouraviev (✉) · A. Koulouri
Dundee Business School, Abertay University, Dundee, UK
e-mail: n.mouraviev@abertay.ac.uk

A. Koulouri
e-mail: a.koulouri@abertay.ac.uk

N. Mouraviev and A. Koulouri (eds.), *Energy Security*,
https://doi.org/10.1007/978-3-030-01033-1_1

in many nations, the existing focus on uninterruptible energy supply does not properly address the core of energy security. Governments often focus their attentions on building supply chains that they perceive as being reliable, based on their political preferences, and to the contracts that might deliver the required volume of energy resources to a country. However, in the modern, turbulent world, supply chains and contracts are subject to considerable power struggles between nations: all kinds of disruptions and renegotiation take place. Fundamentally, supply chains and contracts normally serve the nation's needs in the short and medium terms, rather than the long-term. Within a longer time frame, considering the fact that governments and political regimes change, and the depletion of fossil fuels accelerates, a certain country, particularly one that is resource-poor, will have to continuously deal with its dependency on other nations for the supply of energy resources. Therefore, short- and medium-term security is likely to result in energy insecurity in the long run. Naturally, many nations are concerned with this prospect and are looking for options to remedy the situation. But what are the solutions?

This book discusses the past and current approaches to energy security and offers a novel understanding, which focuses on long-term sustainability. The book investigates energy security from two interrelated perspectives—the increased utilisation of renewable energy sources (RES) and resource use efficiency. In this book, the term *resource efficiency* is used to mean deriving the most value from resource inputs (related to energy production), and incorporates energy efficiency, while materials efficiency is beyond the scope of the book.

These dimensions apply equally to resource-rich and resource-poor economies as all nations are interested in sustainable development, including economic durability, advancement of human capital and a cleaner environment. The focus on renewable energy and resource efficiency addresses all elements of sustainability, and the book argues that these dimensions should be incorporated into a nations' energy policies and governance. Furthermore, the book shows a range of examples, from various countries, of how governments foster renewable energy generation and complement it with programmes and tools that enhance resource use efficiency. To summarise, the book's purpose is to investigate the enablers, policy approaches, governance issues and management problems related to the reduction of dependency on fossil fuels and to highlight the experience of selected economies in implementing resource efficiency programmes. As for resource-rich countries, they often experience path dependency: they keep buying energy sources, while missing out on the opportunity to utilise renewables and increase resource use efficiency. This is why this book should appeal to readers from most nations, rather than just resource-rich countries.

In the context of the increasing significance of renewables, the book examines the following questions:

- What is the relationship between the development of renewable energy technology and energy security?
- Are certain nations able to strengthen their energy security by diversifying the supply of energy resources?
- What are the policy instruments that could be effectively used for the sustainable production of power from RES and for increasing resource efficiency?
- What are the context-specific challenges to energy security in certain countries, such as Australia, Kazakhstan and Ukraine?
- What progressive experience in RES and resource efficiency accumulated by nations across the globe can be borrowed and effectively used in other countries?

This book is non-technical. Rather, it looks at energy and related issues through the lens of policy and governance. Although affordable and technologically advanced solutions are necessary and will naturally form the basis for the promotion of renewables and efficient use of energy resources, often the focus on technology does not produce the required policy results. Having good technological and engineering solutions is just not enough. Each policy has to be supported by organised actors, institutions, tools, funding, procedures and mechanisms (i.e. governance) to ensure successful implementation.

The book's principal argument is that enhancing energy security requires a new approach in policy and governance, which integrates two core components: the emphasis on increasing energy production from renewable sources and resource use efficiency. This notion contrasts with the traditional understanding of energy security as security of supply, which inevitably focuses on the availability of fossil fuels and reinforces the economy's dependency on finite resources. The experience of resource-rich nations, such as Kazakhstan, Russia, Saudi Arabia and the UAE shows that vested interests related to the established oil and gas sector serve as a principal impediment to the use of renewable energy. Despite the growing environmental concerns and the adoption of clean energy policies, e.g. in Kazakhstan and Ukraine, resource-rich economies are lacking in progress in terms of exploiting their renewable energy sources. By offering conceptual chapters and making use of case studies of specific resource efficiency programmes, this book argues that there should be a shift in a nations' energy policy. Governments should adopt a long-term perspective and an approach that focuses on two tasks: renewable energy (i.e. the promotion of increased production and consumption of energy from renewable sources) and resource efficiency (e.g. energy saving measures, government support to the development of resource efficient technology, design of smart grids and deployment of community-level microgrids). In the light of reducing deposits of fossil fuels, in the long run, only this approach will enable nations to secure resources that can meet their energy needs. It is worth noting that energy supply will remain part of energy security as long as a nation uses fossil fuels; however, fossil fuels should be a shrinking part of the nation's energy mix.

By adopting a new concept of energy security, both resource-rich and resource-poor nations need to gradually reduce their dependency on fossil fuels. While many countries are aware of this need, they just disregard it, driven by the short-term considerations, vested interests of the companies engaged in the fossil fuels contracts and imperfections in the legal and governance frameworks. Although the concept of energy security is often used in political rhetoric, in reality not many nations truly focus on renewables and resource use efficiency. Therefore, the book's objective is to emphasise the need to move from rhetoric to policy adoption and to formation of certain governance mechanisms that would ensure policy implementation.

This book draws on many years of the editors' work experience in resource-rich Kazakhstan where the clear majority of people, including practitioners, academics and students, understand energy security as sufficient supply of oil, gas and coal, adequate to meeting the nation's energy needs and provide revenue for the country's economy. Kazakhstan has been using this approach since 1991 and it has led to accelerated resource depletion, although it has not yet led to a restructuring of the economy or a diversifying of the nation's energy mix (i.e. the share of renewables remains negligibly small). Therefore, there is a rapidly emerging need to reconceptualise energy security and revise energy policy. As this book offers insights into a novel concept of energy security, it is likely to be useful for resource-rich nations, such as Kazakhstan, as well as resource-poor nations that need to understand and foresee changes in energy policy and energy mix.

The novel conceptualisation of energy security is a distinctive and original element that distinguishes this book from competing titles. To emphasise this element, the book is divided into two parts. Part I includes conceptual chapters and macro-level studies. Part II discusses practical solutions by providing case studies of government resource efficiency programmes in a range of nations. As Part II discusses pioneering initiatives in resource efficiency, within varying contexts, this explains the choice of nations and their cases. In addition to the survey of the EU initiatives, case studies include Croatia and the UK, which allows the reader to gain a broad and diverse picture of how opportunities in resource efficiency are used in Europe. The chapters/cases demonstrate innovative tools for improving resource use efficiency in different economic conditions and at different levels—the national level (Kazakhstan, Ukraine), regional level (Australia, UK), and the local level (Croatia, UK). The chapter on EU policy and practice regarding resource use efficiency provides a broader overview of major EU initiatives, which is likely to be useful to many readers. No matter what the selection of cases is, it is impossible for any book to capture all available solutions for increasing resource efficiency. With this in mind, the book intends to show best practices that are not as well known about as opposed to the experience of the USA, Germany or Norway about which much literature is available.

The blend of chapters (conceptual plus practice-oriented) reinforces the book's main argument that both renewable energy and resource efficiency should be treated as critical components of energy security. To this end, this edited collection provides a novel outlook and highlights how resource efficiency programmes are designed and implemented at local and regional levels. Academics, students and practitioners may equally benefit from this book by gaining insights into the concept of energy security and learning from international experience of energy efficiency programmes.

Practitioners may find the book useful as it highlights an array of real-life problems that governments and public agencies experience when they form and implement energy policy while addressing energy security concerns. Furthermore, the book offers an assessment of various energy production and consumption scenarios with the purpose to understand the role of renewables in the future energy mix in selected countries and worldwide. The analysis of assumptions, data and conclusions on energy future, followed by critical appraisal, will be of interest to readers who work in the energy sector, including managers and project staff in energy companies across the globe, experts in government agencies involved in energy sector regulation and staff and volunteers in NGOs working on environmental and energy issues. Practitioners may identify commonalities with how energy security is handled in their country: this might be helpful for studying best practice or borrowing from the experience of other nations.

Researchers and students may also benefit from reading this book, as the book presents a concept of energy security that is not yet commonly shared. The book argues that the future of energy mix should be shaped by increasing utilisation of renewables and resource use efficiency. Many countries are struggling with the challenge of renewables and pay little attention to resource efficiency, and by reading this book, academics and students may identify commonalities and differences between the nations in how governance could enhance (or impede) the progress with RES and energy efficiency. Readers will have the opportunity to make a cross-country comparison on a broad range of energy policy issues and draw their own insights from the controversy that surrounds current thinking about how energy security could be achieved. In summary, the discussion of practical issues relating to renewables, the implementation of energy efficiency programmes and the investigation of energy security' theoretical underpinning create a unique blend of practice and theory that may attract a wide spectrum of readers who are interested in energy and sustainability.

Part I

Energy Security: Trends and Policy Challenges

2

Energy Security Through the Lens of Renewable Energy Sources and Resource Efficiency

Anastasia Koulouri and Nikolai Mouraviev

Introduction

This chapter provides a brief overview of the genesis and transformation of the notion of energy security over time, outlines the prevailing shifts in patterns and concepts of energy security in the twenty-first century, and discusses attempts to conceptualise energy security. It also offers a novel conceptualisation of energy security, focusing on increasing the utilisation of renewable energy sources and resource use efficiency.

The need to review the definitions and concepts of energy security stems from at least two reasons: multiplicity of the concepts and their diverse nature. This also raises a question regarding their benefits in terms of utility and contribution to the conceptualisation of this construct.

A. Koulouri (✉) · N. Mouraviev
Dundee Business School, Abertay University, Dundee, UK
e-mail: a.koulouri@abertay.ac.uk

N. Mouraviev
e-mail: n.mouraviev@abertay.ac.uk

N. Mouraviev and A. Koulouri (eds.), *Energy Security*,
https://doi.org/10.1007/978-3-030-01033-1_2

More generally, a question of whether the extant literature provides adequate understanding of energy security requires careful attention.

The chapter begins by analysing how the energy security concept evolved, what the prevailing assumptions were at the end of the twentieth century and how and why they changed later on, in the twenty-first century. The chapter then proceeds to elucidating an emergent paradigm in the understanding of energy security and the reasons behind this phenomenon. Subsequently, numerous attempts to conceptualise energy security are analysed, which is followed by a discussion of principal taxonomies of energy security dimensions. The chapter then offers a novel conceptualisation of energy security, highlights its underlying assumptions, underpinning theoretical framework and benefit.

The Nature and Evolution of the Concept of Energy Security

The Evolution of the Energy Security Conception

In the twentieth century, concerns regarding energy security primarily focused on access to oil. Its geopolitical importance has been recognised since the beginning of the century, not only due to its critical role in the industrialisation of the world economy, but also because of its strategic importance in both World Wars (Yergin 2006; Cherp and Jewell 2011). Post-war, the importance of oil remained high as the primary source of energy for numerous sectors including transport, manufacturing, food production, healthcare and heating and electricity generation (Cherp and Jewell 2011). With a few exceptions, industrialised economies did not produce enough oil to meet their demand and, hence, depended on oil imports from countries that post decolonisation, had become sovereign nations. These countries relied on oil revenue for their economic growth, as well as for strengthening political stability and enhancing society's cohesion.

The established balance in the oil market was shattered in 1973 when most Arab OPEC members and some non-OPEC Arab nations embargoed oil supplies, initially to the USA, the UK, Canada, Japan and the Netherlands, and later on, to other countries in response to the latter's support of Israel during the Yom Kippur War (6–25 October 1973). This led to a sharp increase in oil prices and an ensuing economic crisis. It also resulted in the development of a strategy to make the supply of oil to the industrialised nations less vulnerable to disruptions. A key aspect of this strategy was the establishment of international frameworks. These included: (1) creating the International Energy Agency (IEA) aiming at coordinating the OECD members' response to future disruptions; (2) supporting a global oil market in which no single actor, having concentrated power, has the ability to disrupt supplies; and (3) consolidating the US political influence and projecting its military supremacy in oil-producing regions, articulated by the Carter Doctrine (Carter 1980). In addition, oil extraction began in new areas, such as Alaska and the North Sea, whilst other energy sources, such as gas and nuclear power, and energy efficiency/conservation, received considerable attention, although much attention was still paid to oil.

The strategy had the desired effect during the 1980s and 1990s. Over this period, oil prices significantly decreased, and the main oil producers were 'pro-West' (Kuzemko et al. 2016, p. 163). Furthermore, the market liberalisation paradigm dominated, with those in its favour arguing that the market could regulate itself. This paradigm rested on the premise that a natural equilibrium between supply and demand exists (or should exist) and, hence, viewed energy security as the result of greater reliance on market forces and reduced government intervention. Following this paradigm, energy policy and governance in 1990s included privatisation of the energy companies, liberalisation of the international trade and markets (mainly oil) and establishment of certain international institutions (Kuzemko et al. 2016). Concerns over and, therefore, interest in, energy security were low during this period.

This changed at the beginning of the twenty-first century, when energy security again drew the attention of industrialised nations as well as of rapidly developing ones, such as China and India. The following reasons led to this.

Firstly, the oil price volatility significantly increased. Between 2000 and 2008 oil prices more than quadrupled during a period of sustained growth that was unique in the oil market history (IEA 2013). This was the result of strong market speculation on the prices of crude oil and the effect of Hurricane Catrina on the US oil production (Kuzemko et al. 2016). Following a sharp decrease in price due to the global economic recession of 2008–2009, prices recovered, and a five-year period of stability followed. However, by late 2014, oil prices plummeted again. This was due to a slowdown of growth in emerging economies, such as China, Russia and Brazil. Having experienced rapid growth in the first decade of the twenty-first century, economic development in these nations began to slow down after 2010. It was also due to a failure by OPEC countries to reach an agreement in November 2014 to limit the production of oil.

Secondly, the rapidly growing demand from Asian countries put pressure on oil prices and increased competition for access to supplies. The impact of the emergence of these new powers, that were largely responsible for the growth of demand for energy, was further compounded by the fact that they were not IEA members and, hence, were not bound by its regulations and mechanisms.

Thirdly, there were growing concerns over the availability, accessibility and quality of oil supplies in relation to the current reality of exploiting, or considering the exploitation, of scarcely accessible oil (e.g. shale formations in North Dakota, offshore Arctic gas reserves, Canadian oil sands, deepwater oil off the Brazilian coast). The concerns were about potentially significant increases in costs associated with exploring, extracting and transporting oil and gas from the noted sites, which could lead to much higher energy prices, as well as to serious environmental implications.

Furthermore, the accelerated depletion of oil reserves of OECD nations and other importers has become a growing problem. Consequently, oil production (and supply to many nations across the globe) have been increasingly concentrated in the Middle East and the ex-Soviet space (Cherp and Jewell 2011). This created vulnerability of the energy supply due to events in these countries and, in addition, vulnerability related to the transportation routes from these regions to importers, which are characterised by a number of 'chokepoints' (main ones—in order of importance based on daily transit of volume of oil—the Hormuz Strait, the Malacca Strait and the Suez Canal).

A fifth reason, in contrast to the established paradigm of operating within international frameworks, was due to the formation of state-supported and funded National Oil Companies (NOCs), which created a new dimension of the power struggle in the world oil market. In the case of China (oil importer), NOCs sought to acquire equity in overseas oil and gas production, which has been attained in 20 countries (Bridge and Le Billon 2017). In the case of Russia (oil and gas exporter), NOCs manage the nation's natural resources (including oil and gas) and, hence, play a key role in fostering its economic growth, as well as international economic and political collaboration (Kuzemko et al. 2016).

In essence, the traditionally dominating OECD countries have lost control of both oil supply and demand. The control over supply was lost in the 1960s and 70s with the creation of OPEC and the independence gained by oil-rich nations. The control over demand was lost at the beginning of the twenty-first century with the rapid growth of demand from Asian economies and the formation of NOCs (Bridge and Le Billon 2017).

A number of other factors in the last two decades further aggravated concerns over energy security. Disruptions in the gas supply to the EU nations in 2006, 2009 and 2014 during the Russia-Ukraine crisis led to the realisation of the need to diversify the gas supply routes to Western Europe (Cherp and Jewell 2011; Kuzemko et al. 2016). Furthermore, nuclear accidents (most recently in Fukushima in 2011) have highlighted the vulnerability of energy systems stemming from human errors and natural disasters. This, together with the challenges and ramifications of nuclear waste disposal, has led to significantly curtailing the use of nuclear power by Western countries (Yergin 2011). Furthermore, these accidents, together with the environmental impact and human rights abuses associated with the extraction, processing and use of fossil fuels, gave rise to additional dimensions of energy security concerns relating to sustainability and justice (Dannreuther 2017). Finally, numerous global security challenges have impacted upon the focus of, and attempts to address, energy security concerns including: the emergence of international terrorism, the threat of nuclear weapons proliferation, the circumstances in oil-producing Arab countries (e.g. instability in Libya following the 'Arab Spring', political disputes in post-Saddam Iraq, the impact of sanctions on Iran) and the instability in countries/

regions of Asia (e.g. Afghanistan), Africa (e.g. Niger Delta) and Latin America (e.g. Venezuela, Brazil).

A Shift Towards a Twenty-First-Century Energy Security Paradigm

Although the primary concern regarding energy security over most of the twentieth century was regarding access to oil, the breadth of energy sources, together with technological advancements permitting their exploitation and the multiplicity of threats to and vulnerabilities of energy systems worldwide have led to the broadening of the concept of energy security. This shift was evident from as early as 2005, when the World Bank (2005) defined energy security as '… ensuring countries can sustainably produce and use energy at reasonable cost in order to:

- Facilitate economic growth and, through this, poverty reduction; and
- Directly improve the quality of peoples' lives by broadening access to modern energy services' (World Bank 2005, p. 3).

Further, in 2011 the IEA published the Model of Short-term Energy Security (MOSES) (IEA 2011), which, whilst recognising the continuing importance of oil, highlighted the need for current energy security strategies to consider all energy resources. It further stipulated that strategies should include a comprehensive analysis of political, economic and natural risks associated with them, and of risks related to the energy infrastructure and services. Moreover, whilst the IEA defines energy security on the basis of continuous availability and affordability (IEA 2018), it also recognises that it has many aspects. In the long term, energy security relates to economic development and environmental needs, and, in the short term, it relates to the ability of energy systems to respond promptly and appropriately to abrupt changes in the equilibrium of supply and demand. However, this equilibrium is a matter of domestic price levels in a certain country. A short-term equilibrium might still mean that the price of energy is a major impediment to economic growth as, although it might satisfy consumption in households and firms, it might still be prohibitively high for most investment projects. Therefore, the IEA conceptualisation of energy security lacks precision and also fails to address environmental concerns.

The shift away from a unidimensional concept of energy security is also evident in the EU's energy security strategy, published in 2014 (European Commission 2014). The strategy considers 'a stable and abundant supply of energy' to be critical for the European Union's 'prosperity and security' (European Commission 2014, p. 2). It further includes specific objectives to mitigate *long-term security of supply challenges*. These include: enhancing energy efficiency; moderating demand; developing new technologies; increasing energy production; diversifying suppliers and routes; establishing an internal energy market and missing infrastructure links; strengthening emergency response mechanisms and safeguarding critical infrastructure; and, have *one voice* on external energy policy and improve coordination of national energy policies. Yet, the EU conceptualisation of energy security lacks clarity in failing to define what 'abundant supply' means and to provide an argument for its appropriateness in view of the fact that price serves as a regulatory tool of supply of and demand for all goods and services. Abundance of energy may encourage its inefficient and wasteful utilisation, which is unlikely to align well with energy policy and energy security. Moreover, this definition fails to acknowledge the implications of energy choices for the environment.

A similar direction was taken by the Group of Seven (G7: the United States, Canada, Japan, the United Kingdom, France, Germany and Italy) that in 2014, outlined a set of seven energy security principles, setting a new energy security paradigm, to underpin the national and collective energy security strategies of member states (US Department of Energy 2017). These principles echo—perhaps unsurprisingly—those outlined by the EU in its energy security strategy. The principles require that G7 nations should focus on: (1) developing the energy markets; (2) diversifying energy resources, suppliers and routes, and supporting local energy supply sources; (3) enhancing energy efficiency in demand and supply; (4) investing in research and innovation, and promoting the deployment of clean and sustainable technologies; (5) accelerating the transition to a low-carbon economy and reducing greenhouse emissions;

(6) improving energy system resilience; and (7) developing emergency response systems.

It is worth noting that we are witnessing an emerging shift in the prevailing energy security paradigm, that in the twentieth century focused on securing oil supplies. In general, a policy paradigm is a set of components including: ideas, defined issues, solutions and resources to address them; models; organised actors and their allotted roles; institutions and their capacities and processes; agents with special expertise; and the interaction of all these components (Fosler 1992; Hall 1993; Campbell 2002; Burns and Carson 2009; Carson et al. 2009). Varying opinions exist regarding what set of critical components forms a policy paradigm, although most researchers agree that at least three elements should be included: a concept shared by many (or most) participants; organised actors and specialised institutions; and implementation mechanisms (Mouraviev and Kakabadse 2014). Although a detailed delineation of the evolving twenty-first-century energy security paradigm is beyond the scope of this chapter, a number of elements pointing to its genesis could be highlighted.

Firstly, at the cognitive level, there is general acceptance that energy security concerns have shifted from focusing on access to oil to embracing a multitude of dimensions. Furthermore, analysis of the current energy security definitions and strategies of international institutions, such as the World Bank and the IEA, and supranational organisations, such as the EU and the Group of 7, as well as individual nations, show the emergence of a common set of principles. These include diversification (of sources, suppliers and routes); efficiency (in meeting the demand and in supply chains); sustainability (economic and environmental); system resilience; and emergency response capacity underpinned and facilitated by technological innovation and reliable energy markets. These principles form the conceptual foundations of the emergent paradigm and underpin and inform the energy security strategies and broad policy goals of both industrialised (e.g. Germany, Japan) and developing countries (e.g. China, Brazil).

Secondly, a number of organised actors have emerged, such as the IEA (whose role appears to have evolved from that at its inception, that is, of ensuring undisturbed access to oil for OECD countries), the World Bank, the EU, the World Energy Council, ministries of energy (e.g. Japan's Ministry of Economy, Trade and Industry, METI), environmental groups etc. These are tasked with reconceptualising energy security in view of the dynamic and complex nature of concerns surrounding it, and with designing strategies to mitigate and address them.

Thirdly, the input of agents with special expertise has also been identified with academics and research institutions in the fields of energy policy, renewable energy technologies, natural resource management and environmental sustainability, seeking to enhance our understanding of what constitutes energy security, how best to address concerns surrounding it and how to evaluate the effectiveness of proposed solutions. The growing interest in the field of energy security and numerous other disciplines is demonstrated by the rapidly increasing number of relevant studies and reports in the last 10 years and the engagement of experts by governments to inform policy design and implementation.

A fourth point focuses around the implementation of mechanisms to address the energy security challenges, which have emerged in the nations across the globe in the form of legislative and regulatory acts, such as China's Renewable Energy Law (2005) and Energy Conservation Law (2007), Germany's the Renewable Energy Sources Act (2014), Japan's Feed-in Scheme for Renewable Energy (2010), to name a few. Furthermore, many nations have introduced measures to promote renewable energy on the basis of certain elements: quantity (Renewables Portfolio Standards, RPS), obliging energy utilities to use a minimum amount of RES energy, such as in Canada, the USA, Australia, Sweden, Poland and Romania; price (Feed-in-Tariffs, FiTs), obliging energy utilities to purchase energy from RES at fixed prices, such as in China, Brazil and most of the EU; and a combination of both (mainly RPS, partially FiT), such as in the UK, Italy and Japan.

The ultimate goal of the emergent twenty-first century energy security paradigm is to resolve the energy trilemma that includes energy security, energy equity and environmental sustainability. From a cognitive perspective, having a conceptual framework shared by policymakers, academics and practitioners could enhance the analyses of energy security concerns and the development of policy solutions and implementation mechanisms to address them. Although the formation of a conceptual framework, as well as the creation of specialised institutions, seem to swiftly progress, the implementation tools and their actual application are lagging behind, which confirms an emergent, rather than fully completed, status of the energy security paradigm.

Attempts to Conceptualise Energy Security

Dimensions of Energy Security

Cherp and Jewell (2011) argue that concepts of energy security arose in order to address specific needs, such as fuel supplies for military and transportation uses, uninterrupted electricity supply and effective market operations. They further argue that there are three main perspectives on energy security: (1) *sovereignty* (focusing on threats posed by external actors); (2) *robustness* (focusing on *objective* threats from quantifiable causes); and (3) *resilience* (focusing on unpredictable and uncontrollable threats such as climate variations, regime changes, unforeseeable economic crises). Each of these perspectives and associated solutions are grounded on different fields of knowledge: political science, natural science and engineering, and economics and complex systems analysis, respectively.

Similarly, Sovacool (2011) identified nine energy security dimensions including public policy, diversification, energy services, sustainable development, environmental, climate change, social development, energy poverty, industrial, maritime. He further argued that these reflect the different perspectives that one might adopt in considering energy security, such as geopolitical, scientific, economic, ecological, or social welfare.

Kuzemko et al. (2016) argue that the three most common definitions are *security of supply*; *security of demand* (i.e. demand reliability over the medium to long term, market accessibility, price predictability and protection from the impact of market speculation on energy prices); and *security of energy technologies* (i.e. resilience and technical robustness of the supply system/chain). These meanings of energy security are complemented by additional pieces of analysis, such as economic dimensions to energy security, the role of the energy sector in the economy, the energy intensity of it, pricing, supply and demand, and investment in the energy sector. Kuzemko et al. (2016) also emphasise the impact of what they term as 'non-traditional security issues', such as the interrelations of energy security with water and food security, and the impact of climate change. Similarly to Cherp and Jewell (2011) and Sovacool (2011), Kuzemko et al. (2016) recognise that different actors in varying contexts will have a different standpoint on energy security, be it political, technical, economic or environmental.

There is a further set of distinctions relevant to the application of the concept of energy security in practice, which influence and differentiate how the concept is understood, analysed and utilised (Dannreuther 2017). The first distinction stems from the differences between risks and threats arising from deliberate, intentional acts (e.g. from imposed sanctions or from terrorist attacks) and those that are 'consequences of complex interactions of multiple systems, actors and processes' (Dannreuther 2017, p. 13). In other words, energy security from a political aspect is understood differently as opposed to energy security from technological, legal, regulatory and environmental perspectives.

The second distinction stems from acknowledging that energy security is usually considered in reference to specific energy sources. As such, it is applied in relation to the particularities of the specific energy source arising from its nature and the details of producing, transporting, processing and consuming it. It is also influenced by the source's value as determined by its market price or economic rent (Dannreuther 2017).

The third distinction arises from the conceptualisation of energy security in relation to energy resources, as opposed to energy services, that these resources support. The services supported by energy systems in the industrialised nations have led to increasing prosperity and enhanced well-being. Lack of these services in the developing world has had significant, negative implications for the well-being of individuals affected by very low utilisation of, or lack of access to, energy, use of dirty or polluting fuels and time spent sourcing fuel to meet their basic needs.

Nonetheless, despite any apparent disparities in identified key dimensions and conceptual perspectives of energy security, there is a broad agreement that a wider, integrated approach is required to address the multiplicity of energy security challenges (Yergin 2006; Cherp and Jewell 2011; Kuzemko et al. 2016; Dannreuther 2017). Attempting to address this need, two schools of thought have emerged. The first seeks to address the issue through classification, by clustering energy concerns into themes and creating lists of main energy security elements/ objectives. The second seeks to address the issue through quantification, by developing indices that assess and track the performance of nations in the energy security field using sets of metrics (Cherp and Jewell 2011). The next section reviews the classification methods, whereas a review of the quantification methods is beyond the scope of this chapter.

Energy Security Defined Through Classification of Concerns

In their review of 104 studies of energy security, from 2001 to 2014, including peer-reviewed papers and reports of national agencies, international institutions and business/professional bodies, Ang et al. (2015) identify 83 definitions of energy security and seven themes/dimensions that these definitions focused on or incorporated. These include energy availability, infrastructure, energy prices, environment, societal effects, governance and energy efficiency. Of those, only a handful proposed an energy security definition focusing/based on only one of these dimensions. Moreover, their analysis of how the focus of the proposed definitions evolved over time indicated that energy availability, infrastructure and energy prices were and remain critical concerns. However, over time, energy security has been considered in a more holistic and integrative manner, in recognition of the increasing complexity of the issues it is impacting on and of those that affect it. As such, environmental, societal, governance and energy efficiency dimensions have increasingly been encompassed in energy security conceptions. This evolving pattern of factors considered as critical in conceptualising energy security reflects the changing focus of energy policy and governance concerns at international, regional and national levels.

In another study, Sovacool and Brown (2010), based on their meta-analysis of 91 peer-reviewed papers published between 2003 and

2008, identified four *interconnected factors* on which energy security should be based. These include availability, affordability, energy and economic efficiency, and environmental stewardship. The latter factor incorporates the elements of sustainable use of resources, mitigation and adaptation to climate change and protection of the environment. This study was largely echoed by Ang et al. (2015) who identified similar factors.

Given that both studies—by Sovacool and Brown (2010) and Ang et al. (2015)—were based on the meta-analysis of peer-reviewed papers and, in the case of the latter, various reports, the similarity of the identified themes indicates that the academic community and to a certain extent national and international institutions, have continued to consider the energy security conception in a multidimensional manner, reflecting the complexity of the issues at hand. Moreover, it could be argued that these similarities indicate the emergence of a unified perspective in the conceptualisation of energy security, which might provide a shared language and framework to underpin, develop and evaluate policy and governance efforts to enhance energy security.

Notable taxonomies of energy security concerns/dimensions include:

- The 4As: availability (i.e. physical existence of resources); accessibility (i.e. geopolitical conditions affecting access to resources); affordability (i.e. the cost associated with the full cycle from extraction to consumption); and acceptability (i.e. impact on climate change, environmental degradation, human rights and political stability) (Kruyt et al. 2009);
- Elkind's taxonomy, which similarly to Kruyt et al. (2009), focuses on availability, accessibility, affordability and sustainability (Elkind 2010);
- The '4Rs of energy security' consisting of *review* to understand the issue; *reduce* the amount of energy used; *replace* resources by diversifying; and *restrict* by regulating new demand (Hughes 2009);
- Conceptualisation of energy security by incorporating five dimensions: availability, affordability, technology development, sustainability and regulation (Sovacool and Mukherjee 2011);
- Similar to the above, Sovacool's conceptualisation of energy security, comprising the interrelated elements of availability, affordability, efficiency, sustainability and governance (Sovacool 2013);

- Winzer's simplified definition of energy security as *energy supply continuity*, further divided into *commodity supply continuity*, *service supply continuity* and *continuity of the economy* (Winzer 2012); and
- Johansson's typology of energy security, which distinguishes between the energy system as *an object exposed to security threats*, incorporating the security of supply and the security of demand, and the energy system as *a subject generating or enhancing insecurity*, comprising economic and political, environmental and technological risk factors (Johansson 2013).

The classification of energy security attributes—either based on the meta-analysis of energy security studies or proposed by individual contributions—permits the systematic analysis of energy security challenges and concerns (Dannreuther 2017). Yet, in the case of individual studies, the process and criteria for drawing lists of key characteristics appear arbitrary. Similarly, meta-analysis reports fail to systematically justify the method for selecting energy security issues within their scope and clustering them into key themes (Cherp and Jewell 2011). Furthermore, the classification does not necessarily result in an integrative approach to action based on enhanced understanding and assisting in developing integrated solutions (Cherp and Jewell 2011).

As an alternative to the classification approach, and specifically the influential 4As definition (Kruyt et al. 2009); Cherp and Jewell (2014) adopted the definition of energy security as 'low vulnerability of vital energy systems' (as proposed by, for example, Jewell et al. 2014). Arguing that energy security is 'an instance of security', Cherp and Jewell sought to answer fundamental questions relating to any security concern (as proposed by Baldwin 1997): 'For whom? For which values? From what threats?' (Cherp and Jewell 2014, p. 415). Utilising this approach, they argued, permits the exploration of vulnerabilities through the lens of resilience, exposure to risk, and the interconnection between energy systems and services vital for social welfare and economic development. Furthermore, it offers an integrative approach, drawing from multiple fields of expertise.

Adopting a systems vulnerability analysis perspective permits the comprehensive study of energy systems in their entirety including analysing and assessing the vulnerability of their constituent parts as well as designing mitigating and developmental actions. In contrast, the 4As definition narrowly focuses on the security of supply (Kruyt et al. 2009), omitting key elements and factors of critical importance to the security of an energy system, such as infrastructure and market volatility. Moreover, the systems vulnerability analysis approach allows sufficient flexibility in considering energy systems in diverse contexts. It enables the identification of detailed and contextually specific elements including vital parts of an energy system and areas of vulnerability. It also permits an analysis of potential threats and identification of the system's resilience to them. Furthermore, whilst acknowledging the existence and influence of different contexts and actors, a systems vulnerability analysis facilitates the separation of objectively existing/tangible elements (e.g. energy stocks, infrastructure, markets) from socio-political (subjective) elements; national and/or institutional interests, priorities and objectives; and future expectations.

In summary, a systems vulnerability analysis seeks to draw insights from different disciplines, advance our understanding of energy security and inform policy design, implementation and evaluation. Importantly, it could inform strategies at all levels and underpin the development of implementation mechanisms within the context of the emergent energy security paradigm.

Yet, most recently, Azzuni and Breyer used a vulnerability-focused approach to identify 15 dimensions of energy security: availability, diversity, cost, technology and efficiency, location, timeframe, resilience, environment, health, culture, literacy, employment, policy, military and cybersecurity (Azzuni and Breyer 2018). Their study adopts a conception of energy security as a 'feature (measure, situation, or a status) in which a related system functions optimally and sustainably in all its dimensions, freely from any threats' (Azzuni and Breyer 2018, p. 5). Using this definition as a basis, they identified as many factors as possible with some relationship with energy security arguing that even if a factor is of lesser importance, its inclusion provides additional insights. Although this is undoubtedly true, it is debatable whether incorporating all possible dimensions relating–even tenuously—to energy security promotes our understanding of the concept and, importantly, assists in formulating effective policy and governance to enhance it. Moreover, the energy security definition underpinning the selection of pertinent factors lacks clarity and precision and appears overly general to provide a good grounding for developing, implementing and evaluating energy security policy.

This draw to conceptualising energy security through identifying its constituent parts indicates an implicit or explicit recognition that energy security is a value in continuous tension and competition with other values, such as economic efficiency and growth, sustainability and justice (Dannreuther 2017). As such, it should be considered as an integral part of the *energy trilemma*, that is the search for equilibrium between three fundamental objectives: to ensure energy security, energy equity and environmental sustainability (World Energy Council 2016). In transitioning to secure, affordable and sustainable energy systems, WEC has identified five key strategies. These are: diversifying and transforming the energy supply; expanding infrastructure to improve energy access; pursuing consumer affordability and economic competitiveness; enhancing energy efficiency and managing demand across all sectors of the economy; and decarbonising the energy sector and supporting the shift to a low carbon economy (WEC 2016). WEC's trilemma approach is widely accepted as providing a useful theoretical framework for conceptualising, evaluating and taking action to enhance energy security. However, the associated strategies for resolving, or reaching an equilibrium in, the energy trilemma, provide a set of broad policy goals that require appropriate contextualisation and operationalisation.

The classification of key dimensions of energy security as a means of conceptualising it and addressing the multiplicity of challenges surrounding it offers a basis for a shared understanding amongst energy security scholars, from different disciplines, as well as a shared language for scholars and policymakers. Furthermore, it could underpin international collaboration and enable the formulation of a strategic direction and goals, broadly shared by nations worldwide, to address those challenges and enhance energy security. Nonetheless, identifying key energy security dimensions through the classification of concerns does not remove the need for contextualising these dimensions and, consequently, the strategic direction and goals stemming from them. Moreover, it does not remove the need to operationalise them by designing and implementing appropriate implementation mechanisms. Finally, the utility of such classifications is in direct correlation with their attempt to be all-encompassing and, hence, too general to be of relevance and use in progressing the study of energy security and in informing and underpinning policy formulation and implementation.

Why Reconceptualise Energy Security?

Why is it necessary to revisit the conceptualisation of energy security? As with any study, this need stems from research objectives, the gaps in existing literature, as well as from a new context. Compared to the 1980s and 1990s (and earlier time), the context in many nations, as well as international conditions in energy generation and consumption, have changed dramatically. This was discussed by this chapter in relation to the emergent twenty-first-century energy security paradigm. However, having described its elements, this chapter also challenges the emergent paradigm due to a range of reasons, referring to the gaps in extant literature.

To discuss these gaps, we need to raise the question: what is the foundation for conceptualising (or reconceptualising) energy security? Some answers have already been provided: the foundation might be a nation's specific context; a nation's position in the energy market (e.g. importer or exporter); or a certain perspective (e.g. continuity or vulnerability).

A vast array of available energy security' definitions and concepts display much ambiguity and lack of explanation. For example, what are 'reasonable cost' or the 'affordable price' of energy? What is reasonable and affordable for one, might be unreasonable and too expensive for another. From the economic perspective, the price paid becomes the equilibrium price (and sets the equilibrium quantity), no matter whether it is 'affordable' or 'unaffordable'. At any given time, the price of energy is the one at which quantity demanded equals quantity supplied. Therefore, the use of terminology, such as 'affordable' and 'reasonable' carries little meaning and requires rigorous justification. Similar drawbacks apply to the use of 'abundance' and 'acceptability': whilst abundance might quickly lead to irrational use of energy, acceptability (the impact of a certain national energy system on climate change or political stability) is very difficult to define, measure and monitor, particularly in the long run (Szulecki 2018).

Yet another drawback of the existing conceptualisations of energy security is the tendency towards all-embracing definitions. Including 15 dimensions, such as culture, diversity and others, in energy security (Azzuni and Breyer 2018), inevitably dilutes the focus and makes the definition unclear and indistinguishable from other all-embracing definitions. Whilst these attempts may have their merit as manifestations of a holistic approach, the theoretical underpinnings of these conceptions are difficult to identify.

Furthermore, in contrast to very broad definitions, the narrowly focused definition of energy security as reliable supply omits certain critical dimensions. Is the environment protected? Is security of supply plausible to the majority of the population, from the perspective of energy consumption? Can short-term security of supply provide guarantees for long-term growth? Fundamentally, what theoretical framework might underpin the narrow focus on security of supply?

Energy Security as Ever-Increasing Utilisation of Renewables and Improvement in Energy Efficiency

Prior to discussing our approach to energy security, it is worth stating the assumptions that we made. These are as follows:

- We do not aim to provide a single, universal (one-size-fits-all) definition or concept of energy security. Rather, we acknowledge the significance of three things: (a) a certain field (e.g. political science, technology and innovation or economics of development) that provides the terminology and draws on basic concepts in the field; (b) the context (e.g. national, regional, bilateral, multilateral or international); and (c) diversity of approaches, definitions and concepts (i.e. interdisciplinary and multidisciplinary studies undoubtedly have their merit and should be encouraged).

- We aim for an improved conceptualisation, rather than a novel definition of energy security. This is because we are looking for a rich (or richer) picture of energy and related issues whilst focusing on a definition might result in numerous limitations.
- A theoretical framework that underpins this chapter's understanding of energy security is sustainable development, in its commonly shared meaning of 'development that meets the need of the present without compromising the ability of future generations to meet their own needs' (UNCSD 2001). Focusing on two critical parts of energy security—ever-increasing RES utilisation and efficiency improvements—means pursuing 'opportunities that sustain the natural and/or communal environment as well as provide development gain for others' (Patzelt and Shepherd 2011, p. 632), which are categorised as sustainable development opportunities. Development gain includes economic, environmental and social gains (Leiserowitz et al. 2006; Shepherd and Patzelt 2011; Patzelt and Shepherd 2011) to each of which policy and governance aimed at energy security contribute.
- We view energy security as a critical component of energy policy, i.e. we assume that energy security is, or should be, embedded in policy. In reality, this may or may not be true. It is well known that many nations include mentions of energy security in policy documents setting strategic direction and even set certain goals to achieve energy security. However, implementation tools and mechanisms are often lacking, and energy security remains at the level of high-order political and economic rhetoric, which explains the next point.
- Although we adopted the policy and governance perspective, our focus is on governance, i.e. implementation. This is because the need for effective governance schemes and instruments cannot be overemphasised. This chapter draws on governance as 'a government's ability to make and enforce rules, and to deliver services regardless of whether that government is democratic or not' (Fukuyama 2013, p. 350). Adopting a view that the governance' core is execution (Fukuyama 2013), the chapter does not examine policy agenda of certain nations. Many nations' policies include statements about

energy security, renewables and improvements in the use of energy resources. However, in most countries, progress in achieving set targets has been minimal and/or slow. Rather than examining policy, we aim to identify improvement opportunities related to governance.

- Still, the discussion of governance should not be viewed as normative, as individual contexts are likely to determine individual solutions for each country (see, for example, Chapter 5 about Australia or Chapter 6 about Ukraine). The book offers insights into the governance experience that could be borrowed, adapted, enhanced and used as appropriate in other nations. However, it is unlikely that mechanical copying of experience would work in another context.

The policy and governance perspective inevitably implies a long-term view, as swift changes in the energy sector and, more importantly, quick results are extremely difficult to accomplish. Political approvals, technology adoption, investment decisions and deployment of facilities require considerable time. Although radical policy (or paradigm) shifts are possible, a more realistic approach, based on vast experience across the globe, is the one requiring incremental changes that are consistent and coherent. Therefore, we view energy security as a process: it includes expanding the utilisation of RES and improving resource use efficiency. The term *resource use efficiency (or resource efficiency)* is used from the perspective of deriving the most value from resource inputs (related to energy production), and incorporates energy efficiency, whilst materials efficiency is beyond the book's scope.

The proposed conceptualisation of energy security does not speak directly about security of supply. Of course, a call to disregard continuity and reliability of energy supply would be unwise. From a practical perspective, this call will be premature as the share of renewable energy and savings from efficient energy use are very small in most nations and insufficient to replace fossil fuels to a degree that could be viewed significant. Rather, we argue that nations' energy policies should be complemented by a range of governance tools. Promotion of renewables needs to be in the form of gradual increases of the share of renewable energy in the total energy generation and consumption, which should be supported by instruments, such as government subsidies, low-interest loans, feed-in tariffs and other incentives for producers and consumers. Resource efficiency should also be enabled by implementation mechanisms, such as innovation vouchers, energy-saving measures, incentive schemes and financial support for the development of smart grids and other resource-efficient technology. In addition, governance would benefit from special tools aimed at community mobilisation as local communities show significant potential for both renewable energy generation and improvements in resource use efficiency (e.g. see Chapter 8 about the experience of the south-west of the UK in community mobilisation). To summarise, in both areas—utilisation of RES and resource efficiency—policy objectives (that often are already in place) need to be supported by a broad and elaborate range of governance schemes, procedures and tools.

The proposed approach to energy security may not seem novel as it draws on two items that have been under discussion for decades. However, its novelty is precisely in its focus exclusively on these two parts of energy policy—renewables and efficiency. We view energy security as sustainable use of RES, which means the purposeful and ever-increasing utilisation of renewable sources for the production of power, and a corresponding decrease in reliance on non-renewable sources, complemented by resource efficiency programmes. The latter are a natural part of energy security as efficiency means absence of waste. More efficient use of energy stemming from simple actions, such as better insulation, to sophisticated and technologically challenging solutions, such as smart grids, result in reducing consumption. Although this does not have a direct impact on how (from what sources) energy is generated, optimised consumption might further limit the use of fossil fuels in favour of the enhanced utilisation of renewables. Supported by governance instruments, efficiency measures could be geared not only towards reduction of consumption, but also towards energy efficient solutions aimed at replacing traditional energy use by renewables (e.g. 'green' buildings).

The principal benefit of the proposed approach to energy security can be described as follows. The gradual shift to renewables will take decades, which explains the need not only for long-term policy objectives, but also for governance instruments designed to facilitate and support the lengthy transition from fossil fuels to renewables. In the long run, in the light of reducing deposits of fossil fuels, growing environmental concerns and the pressing need for climate change mitigation, only this approach will enable nations to secure resources that can meet their energy needs. It is worth noting that energy supply will remain part of energy security as long as a nation uses fossil fuels; however, fossil fuels should be a shrinking part of a nation's energy mix. Therefore, although the proposed approach does not directly refer to security of supply, its ultimate benefit is in ensuring reliable and an ever-increasing supply of energy from non-finite energy sources, as opposed to fossil fuels, thus ensuring a nation's sustainable development.

Our conceptualisation of energy security can be well illustrated with the case of Kazakhstan, a resource-rich nation in Central Asia. Kazakhstan's oil sector has been playing a key role in the nation's development to date and can be credited for its contribution to the rapid economic growth following the country's independence from the Soviet Union in 1991. However, Kazakhstan's reliance on oil, gas and coal, due to their role as a principal source of energy and a critical source of budget revenue, has created path dependency. One might argue that the oil lobby constrains the development of renewables due to vested interests. The nation's vast oil reserves provide broad opportunities for the diversification of the economy, although to date, these opportunities have been largely missed. This however raises a concern about the country's sustainability in the long run. At the same time, since 2006, the nation has been taking steps to put in place a policy aimed at promoting renewables, and a large number of laws and regulations have been adopted. Kazakhstan's potential for utilising renewables is significant and estimated at over 1 trillion kWh/year, with power generation from wind and solar resources considered both technically and economically viable (Energy Charter Secretariat 2013; REEEP 2014). Yet, to date, the renewables sector remains underdeveloped, with many elements, such as tariffs, incentives and procedures, yet to be established and institutionalised. Many existing processes and tools that are supposed to incentivise RES promotion are ineffective or unclear for producers and consumers (Koulouri and Mouraviev 2018). In summary, although renewables are part of the country's energy policy, the governance of renewables is yet to be developed. If Kazakhstan is to be successful in reducing its dependency on oil, ensuring energy security and sustainable economic development, it is crucial that the utilisation of renewables significantly expands (Koulouri and Mouraviev 2018). Critically, the ever-increasing utilisation of renewables depends on governance, rather than on policy that is already in place.

Conclusion

Aiming to highlight a range of energy security features, in the twenty-first century, a variety of approaches have led to an emergent paradigm. Many, if not most researchers, view energy security as a multidimensional construct whose conceptualisation is contextual, rather than something that could be universally accepted. The field of study and a certain nation's conditions are important drivers of understanding energy security. An analysis of taxonomies of energy security dimensions shows that there has been a shift from a narrowly focused approach (security means reliable supply of oil) to a much broader, multidimensional conceptualisation. It typically includes management of supply of and demand for energy; development of new technologies; energy production from a range of sources; diversification of suppliers; and resilience of the energy system. These broad approaches to energy security have their merit and, owing to their multiplicity and similarity between them, form the emergent paradigm. However, the broad approaches also carry some drawbacks, amongst which the most critical are lack of precision; an attempt to embrace too many dimensions; failure to address environmental concerns; and insufficient attention to a long term perspective and sustainability.

In contrast to broad views, we offered conceptualisation of energy security that focuses on two interconnected components—ever-increasing power generation from renewables and improvements in resource use efficiency. By emphasising that RES and resource efficiency require effective governance, rather than additional policy acts, the chapter shapes the government's energy security agenda: its nature is in the gradual departure from the dependency on fossil fuels to increasing utilisation of renewables, which would ensure long-term economic, social and environmental sustainability.

The chapter contributes to the debate on conceptualising energy security by offering a clear and focused approach that links governance (of RES and resource efficiency) with sustainability: more renewable energy means less use of fossil fuels; greater resource use efficiency means less demand for energy from all sources; and smaller dependency on non-renewables means greater sustainability. The proposed novel understanding of energy security challenges the emergent paradigm discussed earlier in this chapter. Revisiting the core elements of a policy paradigm (shared ideas; organised actors with a set of responsibilities; institutional capacities for implementation), we argue that a novel approach to energy security, once embedded in governance, is likely to swiftly transform into a policy paradigm. The only element that is currently missing is implementation capacities, i.e. governance. A more detailed discussion of this transformation and conditions under which it could take place might form a useful subject for future research.

References

Ang, B. W., Choong, W. L., & Ng, T. S. (2015). Energy security: Definitions, dimensions and indexes. *Renewable and Sustainable Energy Reviews, 42,* 1077–1093.

Azzuni, A., & Breyer, C. (2018). Definitions and dimensions of energy security: A literature review. *Wiley Interdisciplinary Reviews: Energy and Environment, 7*(1). https://doi.org/10.1002/wene.268.

Baldwin, D. A. (1997). The concept of security. *Review of International Studies, 23*(1), 5–26.

Bridge, G., & Le Billon, P. (2017). *Oil* (2nd ed.). Cambridge: Polity Press.

Burns, T. R., & Carson, M. (2009). Public policy paradigm theory: Paradigm shifts in the EU–social and related policies. In *9th Conference of European Sociological Association on European Society or European Societies?* http://esa.abstractbook.net/abstract.php?aID=3485. Accessed 21 Apr 2017.

Campbell, J. L. (2002). Ideas, politics, and public policy. *Annual Review of Sociology, 28,* 21–38.

Carson, M., Burns, T. R., & Calvo, D. (Eds.). (2009). *Paradigms in public policy: Theory and practice of paradigm shifts in the EU*. Frankfurt am Main: Peter Lang.

Carter, J. (1980, January 23). *State of the union address*. http://www.presidency.ucsb.edu/ws/?pid=33079. Accessed 7 July 2018.

Cherp, A., & Jewell, J. (2011). The three perspectives on energy security: Intellectual history, disciplinary roots and the potential for integration. *Current Opinion in Environmental Sustainability, 3*(4), 202–212.

Cherp, A., & Jewell, J. (2014). The concept of energy security: Beyond the four As. *Energy Policy, 75*, 415–421.

Dannreuther, R. (2017). *Energy security*. Cambridge: Polity Press.

Elkind, J. (2010). Energy security—Call for a broader agenda. In C. Pascual & J. Elkind (Eds.), *Energy security: Economics, politics, strategies and implications* (pp. 119–148). Washington, DC: Brookings Institution Press.

Energy Charter Secretariat. (2013). *Investment climate and market review in the energy sector of Kazakhstan*. www.energycharter.org/what-we-do/investment/investment-climate-and-market-structure/investment-in-kazakhstan-2013/. Accessed 12 June 2018.

European Commission. (2014). *European energy security strategy. Communication from the Commission to the European Parliament and the Council*. COM(2014) 330 final, Brussels, 28.5.2014. https://eur-lex.europa.eu/legal-content/EN/TXT/PDF/?uri=CELEX:52014DC0330&from=EN. Accessed 7 July 2018.

Fosler, S. (1992). State economic policy: The emerging paradigm. *Economic Development Quarterly, 6*(1), 3–13.

Fukuyama, F. (2013). What is governance? *Governance: An International Journal of Policy, Administration, and Institutions, 26*(3), 347–368.

Hall, P. (1993). Policy paradigms, social learning, and the state: The case of economic policymaking in Britain. *Comparative Politics, 25*(3), 275–296.

Hughes, L. (2009). The four 'R's of energy security. *Energy Policy, 37*(6), 2459–2461.

International Energy Agency (IEA). (2011). *Measuring short term energy security*. OECD/IEA. Paris cedex 15: IEA Publications. https://www.iea.org/publications/freepublications/publication/Moses.pdf. Accessed 7 July 2018.

International Energy Agency (IEA). (2013). *World energy outlook 2013*. OECD/IEA. Paris cedex 15: IEA Publications. https://www.iea.org/publications/free-publications/publication/WEO2013.pdf. Accessed 7 July 2018.

International Energy Agency (IEA). (2018). *Energy security*. https://www.iea.org/topics/energysecurity/. Accessed 7 July 2018.

Jewell, J., Cherp, A., & Riahi, K. (2014). Energy security under de-carbonization scenarios: An assessment framework and evaluation under different technology and policy choices. *Energy Policy, 65*, 743–760.

Johansson, B. (2013). A broadened typology on energy and security. *Energy, 53*, 199–205.

Koulouri, A., & Mouraviev, N. (2018). Governance of the clean energy sector in Kazakhstan: Impediments to investment. *International Journal of Technology Intelligence and Planning, 12*(1).

Kruyt, B., van Vuuren, D. P., de Vries, H. J., & Groenenberg, H. (2009). Indicators for energy security. *Energy Policy, 37*(6), 2166–2181.

Kuzemko, C., Keating, M. F., & Goldthau, A. (2016). *The global energy challenge: Environment, development and security*. London: Palgrave.

Leiserowitz, A. A., Kates, R. W., & Parris, T. M. (2006). Sustainability values, attitudes, and behaviors: A review of multinational and global trends. *Annual Review of Environment and Resources, 31*, 413–444.

Mouraviev, N., & Kakabadse, N. (2014). Public-private partnerships in Russia: Dynamics contributing to an emerging policy paradigm. *Policy Studies, 35*(1), 79–96.

Patzelt, H., & Shepherd, D. A. (2011). Recognising opportunities for sustainable development. *Entrepreneurship Theory and Practice, 35*(4), 631–652.

REEEP (Renewable Energy and Energy Efficiency Partnership). (2014). www.reeep.org/kazakhstan-2014. Accessed 27 Jan 2017.

Shepherd, D. A., & Patzelt, H. (2011). The new field of sustainable entrepreneurship: Studying entrepreneurial action linking 'What is to be sustained' with 'What is to be developed'. *Entrepreneurship Theory and Practice, 35*(1), 137–163.

Sovacool, B. K. (2011). *The Routledge handbook of energy security*. Oxon and New York: Routledge International Handbooks.

Sovacool, B. K. (2013). An international assessment of energy security performance. *Ecological Economics, 88*, 148–158.

Sovacool, B. K., & Brown, M. A. (2010). Competing dimensions of energy security: An international perspective. *Annual Review of Environment and Resources, 35*, 77–108.

Sovacool, B. K., & Mukherjee, I. (2011). Conceptualizing and measuring energy security: A synthesized approach. *Energy, 36*(8), 5343–5355.

Szulecki, K. (Ed.). (2018). *Energy security in Europe: Divergent perceptions and policy challenges*. London: Palgrave Macmillan.

UNCSD. (2001). *United Nations commission on sustainable development, Rep. 9th SESSION*. New York: United Nations.
US Department of Energy. (2017, January). *Valuation of energy security for the United States. Report to Congress*. Washington, DC: United States Department of Energy. https://www.energy.gov/sites/prod/files/2017/01/f34/Valuation%20of%20Energy%20Security%20for%20the%20United%20States%20%28Full%20Report%29_1.pdf. Accessed 7 July 2018.
Winzer, C. (2012). Conceptualizing energy security. *Energy Policy, 46*, 36–48.
World Bank. (2005, December 5). *Energy security issues* (No. 36110). Moscow and Washington, DC: The World Bank Group. http://documents.worldbank.org/curated/en/464811468175435408/pdf/361100ENGLISH01gy1Security01PUBLIC1.pdf. Accessed 7 July 2018.
World Energy Council (WEC). (2016). *World energy trilemma 2016—Defining measures to accelerate the energy transition*. In partnership with Oliver Wyman. London: World Energy Council. https://www.worldenergy.org/wp-content/uploads/2016/05/World-Energy-Trilemma_full-report_2016_web.pdf. Accessed 7 July 2018.
Yergin, D. (2006). Ensuring energy security. *Foreign Affairs, 85*(2), 69–82.
Yergin, D. (2011). *The quest: Energy, security, and the remaking of the modern world*. London: The Penguin Press.

3

Increasing Utilisation of Renewable Energy Sources: Comparative Analysis of Scenarios Until 2050

Marat Karatayev, Galiya Movkebayeva and Zhibek Bimagambetova

Introduction

This chapter views energy security through the lens of renewable energy sources (RES) as a major contributor to economic, environmental and social sustainability. Energy security is contextual, dynamic and multidimensional (Ang et al. 2015). This is because its conceptualisation is dependent on and influenced by the context it applies to; it evolves over time, following changes in our knowledge and understanding of the natural world and technological advances, and integrates numerous dimensions related to resource availability and accessibility, technological capacity, policy, governance and sustainable development. Energy

M. Karatayev (✉)
Energy Technologies Research Institute, University of Nottingham, Nottingham, UK
e-mail: marat.karatayev@nottingham.ac.uk

G. Movkebayeva · Z. Bimagambetova
School of International Studies, Al-Farabi Kazakh National University, Almaty, Kazakhstan

N. Mouraviev and A. Koulouri (eds.), *Energy Security*,
https://doi.org/10.1007/978-3-030-01033-1_3

security has increasingly been taking a central place on the policy agenda of many developed and developing nations. Meanwhile, there is an emerging consensus on the seriousness of the impact of climate change and the need for mitigating action. In view of this, the recent Paris Agreement has reinforced the requirement for a challenging low-carbon transformation of the global energy sector (UN 2015).

Regarding the link between renewables and energy security, it is recognised that RES have the potential to become one of the principal contributors supporting the energy security and sustainable development of both industrialised and developing countries (IPCC 2007; Ölz et al. 2007). RES are inexhaustible, contrary to fossil fuels. An abundant supply of solar, wind, hydro, biomass and geothermal resources is available in many regions worldwide with the potential to provide for over 10,000 times the current global energy needs. This is significantly more than other energy sources, such as geothermic or tidal energy, nuclear power and fossil fuel burning (Ellabban et al. 2014). Increasing the share of RES in the energy mix will gradually reduce nations' dependence on fossil fuels and their vulnerability to the impacts of price volatility, which characterises the energy commodities markets (Francés et al. 2013). Furthermore, RES infrastructure could be of a smaller scale than that of traditional energy, thus making it ideal for decentralised power generation in rural areas (Bassam 2001). Finally, RES energy production mitigates against the environmental impact of traditional energy generation, protecting the environment, local inhabitants, and their living conditions (Panwar et al. 2011).

RES are often studied in relation to their environmental impact and climate change (IPCC 2007), whilst little attention is paid to their potential contribution to nations' energy security. This chapter aims to use energy security indicators to examine the relationship between the deployment of renewable energy technologies and energy security in current and future scenarios. These indicators include the Total Primary Energy Supply (TPES[1]), the share of RES in TPES, the energy

[1]TPES: Total Primary Energy Supply is made up of the indigenous production plus imports minus exports, international marine and aviation bunkers, and plus/minus stock changes (International Energy Agency, https://www.iea.org/statistics/resources/balancedefinitions/#tpes, accessed on 05/12/2016).

import dependency, the energy intensity, electricity production technologies, energy-related carbon dioxide emissions, carbon emissions trading prices and energy prices. In particular, it aims to provide insights into the future energy consumption and supply mix; the depletion rate of natural resources; the technical and economic potential of renewable energy; the fluctuations/adjustments of energy and carbon emissions trading prices; and the plausible changes of the levels of energy-related emissions to the atmosphere.

The chapter begins with an overview of the criteria used for selecting the quantitative and qualitative scenario studies analysed, providing a brief summary of their elements and underpinning assumptions. It then discusses key drivers shaping the global energy supply and demand. Furthermore, it considers the plausible transformative impact of changes to the key drivers on different sectors of the energy system, within a time horizon of 2050. The chapter next discusses sources of uncertainty likely to affect the realisation of the RES' potential in the future. It finally concludes by summarising key learning points emerging from the scenario analysis and providing recommendations for developing strategies to enhance energy security.

Selected Scenarios

This study is based on the review and analysis of quantitative and qualitative scenario studies. Scenario studies are recognised as a useful approach to widen the focus of short-term policy debates, and to explore more radical system changes (Söderholm et al. 2011). Scenario studies can explore technologies, costs, policies, investments, emissions, social appropriateness and shares relative to fossil fuels and nuclear energy (Martinot et al. 2007). International governmental and non-governmental organisations design specific baseline and policy scenarios to predict the effects of different policies on future energy security performance at global and regional levels and their likely impact on energy security indicators.

In this study, energy scenarios were selected on the basis of the following criteria: a time horizon between 2010 and 2050; quantitative and qualitative analysis; assumptions on technical, demographic, economic, social, and political parameters; focus on renewable energy technologies and nuclear energy; geographic coverage of the Organisation for Economic Cooperation and Development (OECD) countries, China, India and other Asian non-OECD member countries. Furthermore, the selected scenarios were released in the last 7 years (between 2010 and 2017); are of international relevance; and in their totality, cover a broad range of stakeholder groups.

Using the above criteria, the following studies were selected:

- World Energy Outlook 2012 (IEA 2012);
- Energy Technology Perspectives—Pathways to a Clean Energy System to 2050 (IEA 2012);
- Energy Revolution: A Sustainable World Energy Outlook (EREC et al. 2012);
- The Energy Report: 100% Renewable Energy by 2050 (WWF and ECOFYS 2011);
- Global Energy Assessment: Towards a Sustainable Future (IIASA 2012);
- BP Energy Outlook 2030 (BP 2013);
- The Outlook for Energy: A View to 2040 (ExxonMobil 2012);
- Global Wind Energy Outlook (GWEC 2013);
- International Energy Outlook 2011 (US EIA 2011);
- Future World Energy Scenarios (Enerdata 2017).

Table 3.1 provides an overview of the selected scenarios with details on their time horizon and key characteristics.

Key System Drivers

This section focuses on key system drivers that shape global energy supply and demand. There are a number of drivers contributing to significant increases in energy demand. These are population dynamics and urbanisation, economic development and GDP growth, degree of policy action and geopolitical relationships, technical and structural

Table 3.1 Overview of analysed scenarios

Scenario title	Time horizon	Principal features
World energy outlook 2012 (IEA 2012)	2035	Three main scenarios were considered: the Current Policies Scenario (CPS); the New Policies Scenario (NPS); and the 450 parts per million of carbon dioxide equivalent (ppm CO_2eq) scenario. The CPS scenario takes into account policies announced and in place by various countries in the world. The NPS scenario takes into account long term energy and climate objectives and policies that have yet to be implemented. The 450 scenario has a different perspective, setting the target of a 2 °C rise of the global temperature and discussing policies necessary to decrease greenhouse gas emissions below current levels
Energy technology perspectives—pathways to a clean energy system to 2050 (IEA 2012)	2050	Three possible energy futures are considered, the boundaries of which are set by total energy-related CO_2 emissions: (a) The 6 °C Scenario (6DS), an extension of current trends that assumes no new policy action is taken to address climate change and energy security concerns and that results in an average global temperature rise of at least 6 °C in the long term; (b) The 4 °C Scenario (4DS), which represents a concerted effort to move away from current trends and technologies, with the goal of reducing both energy demand and emissions vis-a-vis the 6DS; (c) The 2 °C Scenario (2DS), which describes a vision of a sustainable energy system and emissions trajectory that recent climate science research indicates would give an 80% chance of limiting average global temperature increase to 2 °C
Energy revolution: a sustainable world energy outlook (EREC et al. 2012)	2050	Two scenarios are considered: Reference scenario and Energy Advanced [R]evolution Scenario. The Reference Scenario takes into account international energy and environmental policies worldwide. The Energy Advanced Revolution scenario (E[R]evolution) looks at the potential of reducing energy demand through energy efficiency and providing electricity through decentralised renewable energy sources. It explores the use of smart grids to connect these decentralised systems and assumes that hybrid electric cars will be predominant in 2050 and that nuclear energy will be phased out by 2050

(continued)

Table 3.1 (continued)

Scenario title	Time horizon	Principal features
The energy report: 100% renewable energy by 2050 (WWF and ECOFYS 2011)	2050	The study covers all countries and assesses the technical feasibility of achieving 100% renewable energy generation from 2010 up to 2050. The 100% renewables scenario assumes that all countries make extensive use of their renewable energy potentials and establish large-scale electricity storage facilities, smart electricity transmission grids and demand-side management
Global energy assessment (gea): towards a sustainable future (IIASA 2012)	2050	The report develops 60 alternative pathways through which energy systems could be transformed to address critical energy and environmental goals: stabilising global climate change to global temperature of 2 °C above pre-industrial levels; enhancing energy security through diversification of the energy supply; and eliminating household and ambient air pollution. The pathways are grouped into three different approaches for achieving these goals: GEA-Supply, GEA-Mix and GEA-Efficiency. GEA-Supply considers futures with radical developments such as hydrogen or carbon capture and storage. GEA-Mix relies more on today's advanced infrastructure such as biofuels. GEA-Efficiency relies more on today's advanced options such as efficiency and RES
BP energy outlook 2030 (BP 2013)	2030	The report encompasses two sets of data: a base case and a policy case. In the base case, population and income growth are assumed to be the major driving force of energy demand, whereas in the policy case, some stringent policy measures are adopted to cut the CO_2 emissions of developed countries and to reduce the energy intensity of developing countries
The outlook for energy: a view to 2040 (ExxonMobil 2012)	2040	The report offers ExxonMobil's vision for future energy trends. It seeks to answer the following questions: what types of energy will the world use, and how much? How will patterns of demand and sources of supply evolve in different countries? And how will new technologies affect the energy mix and overall energy efficiency?

(continued)

Table 3.1 (continued)

Scenario title	Time horizon	Principal features
Global wind energy outlook (GWEC 2013)	2050	The study examines the future potential of wind power up to 2050 and is based on three distinct scenarios: The Reference scenario, the Moderate scenario, and the Advanced scenario. The Reference scenario adopts the assumptions of the IEA World Energy Outlook, with restrained progress in renewable energy and wind power. The Moderate scenario assumes that current renewables and CO_2 targets will be met. The Advanced scenario explores the implications of achieving an ambitious plan to unleash the full potential of the wind industry
International energy outlook 2011 (US EIA 2011)	2050	The report is based on the US Energy Information Agency (EIA)'s outlook assessment of energy markets, focusing mainly on marketed energy. Projections are generated from the EIA's World Energy Projection Plus (WEPS+) model, which takes into account population growth, economic growth, energy intensity, and historical energy market data. It does not reflect the potential impact of proposed legislation, regulations, or measures
Future world energy scenarios (Enerdata 2017)	2040	The report presents three global energy scenarios developed by the POLES world energy-economy simulation model. The Ener-Brown scenario assumes the gradual improvement on energy intensity; high growth in developing countries (74% of the demand growth over 2015–2040 will come from Asia, and 15% from Africa); fossil fuels renaissance; lower energy prices; diffusion of unconventional energy sources; and continued efforts on renewables. The Ener-Green scenario considers ambitious energy efficiency policies; regular updates of efficiency targets; fossil fuel subsidies phase-out; strong development of renewables (RES+ nuclear development: ~ 70% of power capacities by 2040); and one kWh generated will produce 75% less CO_2 emission in 2040 vs 2015. The Ener-Blue scenario assumes tensions on available resources; increasing energy prices; and diversification towards renewables

Source Compiled by the authors

change, technology costs, fossil fuel prices and CO_2 emission pathways and allowance costs.

Demographic Change

Future population development is an important factor in energy scenario building because population size affects the size and composition of energy demand, directly and indirectly, through its impact on economic growth and development. Most of the scenarios use the United Nations Development Programme (UNDP) projections for population growth as shown in Table 3.2. Based on the UNDP's 2010 assessment, the world's population is expected to grow by 0.76% on average over the period 2010–2050, from 6.8 billion people in 2010 to nearly 9.4 billion by 2050 (UNDP 2010). From a regional perspective, the population of the developing regions will continue to grow most rapidly. The population of Eastern Europe/Eurasia will face a continuous decline, followed after a short while by the OECD Asia Oceania. The populations in OECD Europe and in OECD North America are expected to increase through 2050. Meanwhile, the share of the population living in today's non-OECD countries will increase from the current 82–85% in 2050. China's contribution to the world population will drop from 20% today to 14% in 2050, whilst Africa will remain the region with

Table 3.2 Population projections by regions (in millions)

Region	2010	2015	2020	2025	2030	2040	2050
World	6818	7284	7668	8036	8372	8978	9469
OECD Europe	555	570	579	587	593	599	600
OECD North America	458	484	504	524	541	571	595
OECD Asia and Oceania	201	204	205	205	204	199	193
Eastern Europe	339	340	341	340	337	331	324
India	1208	1308	1387	1459	1523	1627	1692
China	1342	1377	1407	1436	1452	1474	1468
Non-OECD Asia	1046	1128	1194	1254	1307	1392	1445
Latin America	468	499	522	544	562	589	603
Africa and Middle East	1202	1274	1528	1687	1857	2226	2450

Source Adapted by the authors from the UN world population prospects—2010 revision (UNDP 2010)

the highest growth rate, with a share of 24% of the world population in 2050. This projected population growth will lead to an increased demand for energy.

Table 3.2 provides an overview of projected population growth in different regions of the world from 2020 to 2050.

Urbanisation

Increasing urbanisation is the second critical long-term demographic feature, frequently ignored in energy studies. More than 80% of the population of industrialised countries lives in urban areas, and many developing countries show similar high urbanisation rates. According to the UN's last estimate (UN 2014), 4.3 (60.6%) of 7.1 billion people worldwide live in urban agglomerations and, over the next decade, the urban population is projected to increase to almost 5.2 billion. An increasing percentage of this urban population will live in megacities with over 10 million inhabitants. Moreover, it is estimated that eight cities will have more than 15 million inhabitants each; only two of these, Tokyo and New York, are in highly industrialised countries. The remaining six (Beijing, Mumbai, Kolkata, Mexico City, Sao Paolo and Shanghai) are in developing countries. Providing adequate and clean energy services for a world in which the population lives predominantly in urban areas will be a challenging task, due to infrastructure (i.e. capital) requirements, enormous spatial energy demand densities and the need for clean energy in order to mitigate the creation of urban smog from coal fires and/or dense motorised traffic that currently plagues most megacities.

Macroeconomic Development

Economic growth is a key driver for energy demand. Since 1971, each 1% increase in global Gross Domestic Product (GDP) has been accompanied by a 0.6% increase in primary energy consumption (IEA 2012). The decoupling of energy demand and GDP growth is therefore a prerequisite for all scenarios as rapid economic development is propelling significant increases in energy use. GDP growth in all regions is expected to slow gradually over the coming decades (EREC et al. 2012). World GDP is assumed to grow on average by 2.2% per year over the period 2035–2050, compared to 3.1% from 1971 to 2010 (Table 3.3). China and India are expected to keep a higher GDP growth than other regions, followed by the Middle East, Africa, and the rest of non-OECD Asia. GDP in OECD Europe and in OECD Asia Oceania is assumed to grow by around 1.6 and 1.3% per year, respectively, over the projection period, whilst economic growth in OECD North America is expected to be slightly higher. Table 3.3 provides an overview of projected GDP development to 2050.

Development and Geopolitics

Examining demographics in relation to economic development, it becomes clear that there will be a long-term shift in the geographical focus of energy use. In 2010, developing countries' share of energy use was 40%. By 2050, the share of developing countries is estimated to range between 60 and 70%, i.e. a complete reversal of the current energy geopolitical situation. In fact, most scenarios (e.g. IEA 2012; Greenpeace 2012; BP 2013; ExxonMobil 2012) project faster energy demand growth in developing countries, particularly in China, India, Pakistan, Indonesia, Malaysia and countries in Latin America, the

Table 3.3 GDP development projections (in %)

Region	2010–2020	2020–2035	2035–2050	2010–2050
World	4.2	3.2	2.2	3.1
OECD North America	2.7	2.3	1.2	2.0
OECD Asia and Oceania	2.4	1.4	0.5	1.3
OECD Europe	2.1	1.8	1.0	1.6
India	7.6	5.8	3.1	5.3
China	8.2	4.2	2.7	4.7
Non-OECD Asia	5.2	3.2	2.6	3.5
Latin America	4.0	2.8	2.2	3.5
Middle East	4.3	3.7	2.8	3.5
Africa	4.5	4.4	4.2	4.4

Source Adapted by the authors from EREC et al. (2012)

Middle East and Africa. These trends will have an impact and lead to changes in energy systems and energy geopolitics (Umbach 2010; Lombardi and Gruenig 2016).

Infrastructure

Despite energy geopolitics, market exclusion remains a serious challenge. To date, some two billion people do not have access to modern energy services due to a lack of energy infrastructure. Many regions are overly dependent on a single, locally available resource, such as traditional fuelwood or coal, and have limited access to the clean flexible energy forms required for economic and social development (IEA 2012). Moreover, urban and rural poor populations need to get connected to energy grids in order to have access to modern energy services. Infrastructure is the backbone of the energy system, and the future requirements for new infrastructure will be truly vast (IEA 2012). New decentralised energy options could help to reduce energy costs in rural areas, with high cost decreasing through research and development efforts and experience gained in niche market applications. Meanwhile, improving interconnections of energy grids for natural gas and electricity on a continental scale remains a task ahead for many regions, in particular Asia, Latin America, and—in the long term—Africa. Moreover, development of the energy infrastructure is required in Eurasia, particularly to match the abundantly available resources of oil and gas in the Caspian region, Siberia and the Arctic Area with the newly emerging centres of energy consumption in Asia.

Technical and Structural Change

New energy technologies on the supply side and for energy-using devices are expected to lie at the heart of renewable energy development, efficiency improvements, cost reductions and better services provided by the energy sector. In high economic growth scenarios where large quantities of fuels will be required, there need to

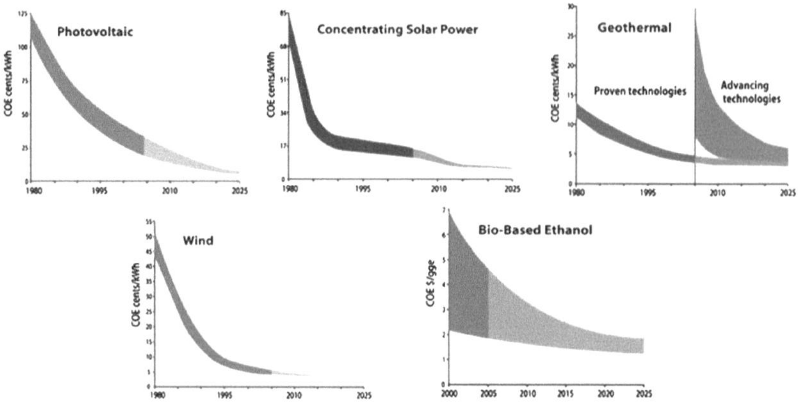

Fig. 3.1 Forecasted renewable energy technology cost (*Source* Adapted by the authors from NREL 2015)

be considerable advances in hydrocarbon exploration and extraction, renewable and nuclear electricity generation, hydrogen and biofuel production and conversion, and more efficient and smart grid systems. In ecologically driven scenarios, low-carbon fossil and renewable technologies are favored and renewable energy technologies costs—indicated as light areas on Fig. 3.1—are projected to decrease by 2025. In fact, the cost of photovoltaic panels, wind and solar power technologies and biotechnologies—shown as dark areas on Fig. 3.1—has dropped significantly over the last 20 years, and it is projected to continue decreasing in the future.

Figure 3.1 provides an overview of the projected cost of photovoltaic, concentrating solar power, geothermal, wind and bio-based ethanol technologies up to 2025.

Fossil Fuel Prices

The projections of energy prices are a very uncertain issue, as fuel prices will continue to be set by market supply, demand and regulatory dynamics. The recent increase of shale gas and shale oil production as well as the increase of renewable energy use has resulted in low price projections for fossil fuels. For example, under the 2014 oil and gas price scenario from Bloomberg, an oil price of just 45–50 USD per barrel was assumed for 2030 (Bloomberg 2015). According to Future World Energy Scenarios, based on supply-demand fundamentals, long-term oil prices should move in a range of 60–100 USD per barrel (Enerdata 2017). However, the projections in the WEO Current Policies scenario and Greenpeace scenario might still be considered too conservative; both scenarios have assumed a price development path for fossil fuels significantly higher than the Bloomberg scenario as can be seen in Table 3.4.

Table 3.4 provides projections of crude oil and natural gas imports and biomass prices up to 2050.

Emission Pathways

Over the last decade, CO_2 emissions have increased by an average of 3% per year, despite the increased focus on climate change and its mitigation. This was mainly the result of high economic growth, particularly in coal-based economies, and higher oil and gas prices which led to an increase in coal-fired power generation. In the future, emissions are expected to grow further. In the WEO 2012 Reference scenario, CO_2 emissions are expected to increase from 29 Gt (gigatons) CO_2 in 2007 to 40 Gt by 2030. CO_2 emissions continue to grow in the ETP 2012 Baseline scenario projections beyond 2030, reaching 57 Gt in 2050. Nearly all the growth in global CO_2 emissions in the Baseline scenario comes from outside the OECD; emissions from non-OECD countries are expected to grow from 15 Gt CO_2 in 2007 to 42 Gt CO_2 in 2050, whilst OECD emissions grow from 14 Gt CO_2 to 15 Gt CO_2 over the same period. Most of the increase in the OECD countries comes after 2030 and long-term emission projections are highly uncertain. In the WEO 2012 higher GDP case, CO_2 emissions reach 43 Gt by 2030, compared to 40 Gt in the Reference scenario and 38 Gt in the low GDP case. Similarly, the high energy demand projections for 2050 indicate that emissions could be up to 20% higher than the 57 Gt projected in the Baseline scenario for that date.

Table 3.4 Development projections for fossil fuel and biomass prices in USD 2010

Fossil fuel	Unit	2010	2020	2025	2030	2035	2040	2050
Crude Oil imports								
Historic price (WEO)		78						
WEO 450 ppm scenario	Barrel	78	97	97	97	97		
WEO current policy	Barrel	78	106	106	135	140		
Energy revolution (2012)	Barrel	78	112	112	152	152	152	152
Natural gas imports								
Historic price (WEO)								
USA	GJ	4.64						
Europe	GJ	7.91						
Japan	GJ	11.61						
WEO 450 ppm scenario								
USA	GJ	4.64	6.86	8.44	8.85	8.23		
Europe	GJ	7.91	10.34	10.34	10.23	9.92		
Japan	GJ	11.61	12.66	12.66	12.77	12.77		
WEO current policy								
USA	GJ	4.64	7.39	8.12	8.86	9.50		
Europe	GJ	7.91	11.61	12.56	13.29	13.72		
Japan	GJ	11.61	14.24	14.98	15.61	16.04		
Energy revolution (2012)								
USA	GJ	4.64	10.84	12.56	14.57	16.45	18.34	24.04
Europe	GJ	7.91	16.78	18.22	19.54	20.91	22.29	26.37
Japan	GJ	11.61	19.08	20.63	22.12	23.62	25.12	29.77
Biomass								
Energy revolution (2012)								
OECD Europe	GJ	7.80	9.32	9.72	10.13	10.28	10.43	10.64
OECD North America	GJ	3.44	3.85	4.10	4.36	4.56	4.76	5.27
Other regions	GJ	2.84	3.55	3.80	4.05	4.36	4.66	4.96

Source Adapted by the authors from WEO (2012), OECD and IEA (2012), EREC et al. (2012)

Table 3.5 The assumptions on CO_2 emissions cost development, in USD

Countries	2020	2030	2040	2050
Annex-B countries	25	40	55	75
Non-Annex-B countries	0	40	55	75

Source Adapted by the authors from UNFCCC (2011)

Cost of Carbon Emissions

Assuming that a carbon emissions trading system is established across all world regions in the longer term, the cost of CO_2 allowances needs to be included in the calculation of electricity generation costs. All analysed scenarios consider CO_2 emissions cost development as applied in the Kyoto Protocol for countries in Annex B (in which 37 countries have binding targets including Australia, European Union member countries, Belarus, Iceland, Kazakhstan, Lichtenstein, Norway, Russia, Switzerland and Ukraine) and non-Annex B countries without binding targets (including Argentina, Azerbaijan, Brazil, Chile, China, India, Iraq, Iran and Mexico) (Table 3.5).

Table 3.5 provides projected CO_2 emissions costs up to 2050 for Annex B and Non-Annex B countries of the Kyoto Protocol.

Plausible Transformations of the Energy System: Exploring the Impact of Scenarios on Sectors

This section discusses possible transformations of energy systems up to 2050 from a sectoral perspective. In particular, the section focuses on power generation, nuclear power and the transport and construction sectors. In addition, it considers plausible trajectories of carbon emissions reduction and the cost of transition to low-carbon energy.

Power Generation

World Energy Outlook (WEO 2012) develops three main scenarios: the Current Policies Scenario (CPS), the New Policies Scenario (NPS), and the 450 parts per million of carbon dioxide equivalent (ppm CO_2eq) scenario. The NPS details the impact of existing policy commitments and the implementation of those recently announced on the key energy demand, supply, trade, investment and emissions trends in the period up to 2035. In this scenario, fossil fuel subsidies are assumed to be phased out by 2020 in all net energy importing countries and more gradually in exporting ones that have announced plans to do so. Furthermore, global primary energy demand is projected to increase by 35% between 2010 and 2035. Approximately 90% of that increase will come from non-OECD countries as the share of OECD countries in world energy demand is projected to fall to about 35% by 2035. Oil, coal, and gas will thus remain the predominant sources of energy (Fig. 3.2). Nevertheless, renewable energy sources will continue to grow; from 13% of the global primary energy demand to 18% at the end of the outlook period, whilst the share of nuclear power will remain constant at around 6–7%.

Electricity demand is projected to increase by more than 70%, with over 9300 GW of installed capacity needed by 2035. Although the share of fossil fuels in electricity generation is expected to decrease to approximately less than 10% compared to 2010, fossil fuels will still provide more than 60% of the electricity supply in 2035 (Fig. 3.3). Electricity generated from renewables is projected to almost triple, and their share in the world electricity mix is expected to increase from 20 to 31%. Hydro energy provides and will continue to provide the largest share of renewables. The share of nuclear power is projected to remain

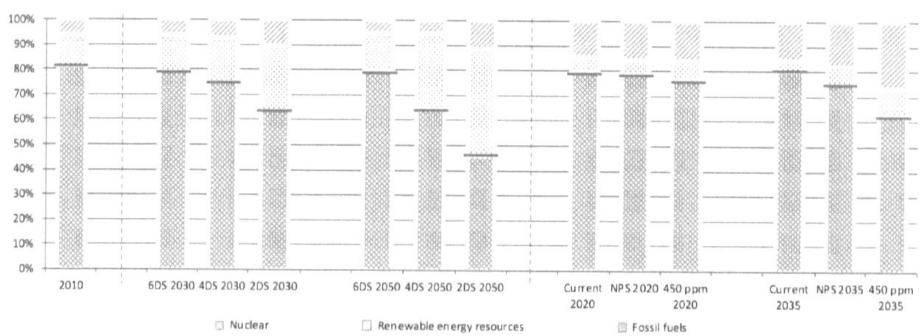

Fig. 3.2 Share of primary energy sources, in the energy demand worldwide, in % (*Source* Compiled by the authors)

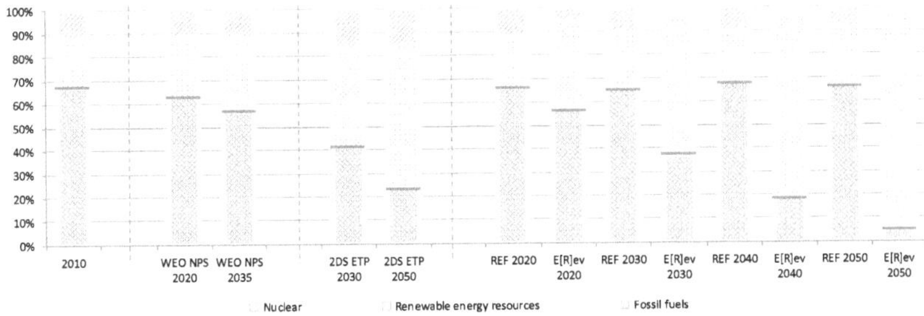

Fig. 3.3 Share of primary energy sources, in the electricity generation worldwide, in % (*Source* Compiled by the authors)

constant at around 12–13%, whilst wind power will have a growing role with its share in the electricity mix increasing to 7.3% in 2035, up from 1.6% in 2010. The contribution of biomass and solar energy technologies in power generation will also increase.

Figure 3.2 provides a comparative overview of the share of primary energy sources in worldwide energy demand between 2010 and (1) 2020 and 2035, for the three scenarios of the World Energy Outlook (WEO 2012) (CPS, NPS and 450 ppm); and (2) 2020 and 2050, for the three scenarios of the Energy Technology Perspective Report (IEA 2012) (2DS, 4DS and 6DS).

Figure 3.3 provides a comparative overview of the share of primary energy sources in electricity generation worldwide between 2010 and (1) 2020 and 2035, for the World Energy Outlook (2012) NPS scenario (WEO NPS); (2) 2030 and 2050, for the Energy Technology Perspective Report (IEA 2012) 2DS scenario (2DS ETP); and, (3) 2020, 2030, 2040 and 2050, for the two Energy Outlook scenarios (EREC et al. 2012), the Reference Scenario (REF) and the Energy Advanced Revolution Scenario (E[R]ev).

The Energy Technology Perspective Report (IEA 2012) develops three possible energy futures, the boundaries of which are set by total energy-related CO_2 emissions:

1. The 6 °C Scenario (6DS), an extension of current trends that assumes no new policy action is taken to address climate change and energy

security concerns, resulting in an average global temperature rise of at least 6 °C in the long term.
2. The 4 °C Scenario (4DS), which represents a concerted effort to move away from current trends and technologies, with the goal of reducing both energy demand and emissions vis-a-vis the 6DS.
3. The 2 °C Scenario (2DS), which describes a vision of a sustainable energy system and emissions trajectory that recent climate science research indicates would give an 80% chance of limiting the average global temperature increase to 2 °C.

The 2DS scenario reflects a concerted effort to reduce overall consumption and replace fossil fuels with a mix of renewable energy resources, and dramatic improvements in terms of energy efficiency. In the 2DS, Total Primary Energy Supply increases by approximately 37% between 2010 and 2050. This is significantly lower than the 85% rise in the 6DS. In the OECD, TPES is projected to stay almost constant in the 2DS and increase only moderately in the 6DS. In the non-OECD countries, even in the 2DS, TPES is projected to rise by about 70% from 2010 to 2050. The share of renewables in the world energy mix will be 43% in the 2DS, 30% in the 4DS and 15% in the 6DS in 2050, compared to 13% in 2010 (Fig. 3.4). In all scenarios, fossil fuels remain a significant part of the global energy system through 2050.

Figure 3.4 provides a comparative overview of the share of primary energy sources in the energy demand worldwide, between 2010 and (1) 2020, 2030, 2040 and 2050, for the two Energy Outlook scenarios (EREC et al. 2012) (the Reference Scenario (REF) and the Energy Advanced Revolution Scenario [E[R]ev]); (2) 2020, 2030, 2040 and 2050, for the Energy Report (WWF and ECOFYS 2011); and (3) 2025 and 2040, for the ExxonMobil (2012) future energy vision.

Figure 3.5 provides a comparative overview of the share of primary energy sources in electricity generation worldwide between 2010 and (1) 2020, 2030, 2040 and 2050, for the two Energy Outlook scenarios (EREC et al. 2012) (the Reference Scenario (REF) and the Energy Advanced Revolution Scenario [E[R]ev]); (2) 2020, 2030, 2040 and 2050 for the Energy Report (WWF and ECOFYS 2011); and, (3) 2025 and 2040, for the ExxonMobil (2012) future energy vision.

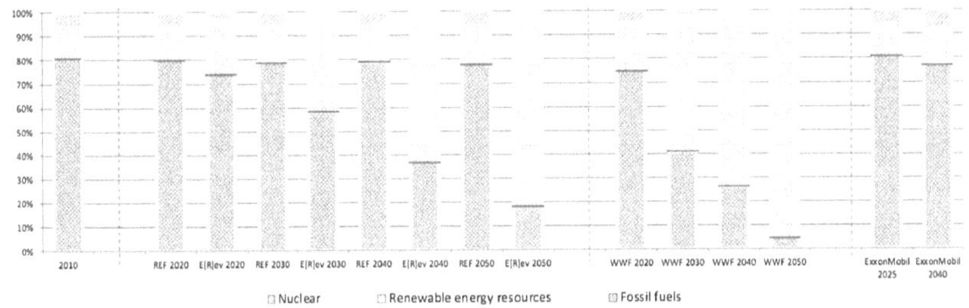

Fig. 3.4 Share of primary energy sources, in the energy demand worldwide, in % (*Source* Compiled by the authors)

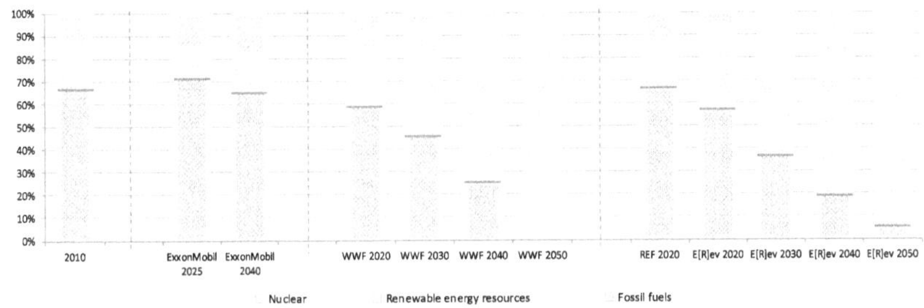

Fig. 3.5 Share of primary energy sources, in the electricity generated worldwide, in % (*Source* Compiled by the authors)

The Energy Outlook (EREC et al. 2012) uses a bottom-up, technology-driven approach to illustrate the possibility of 100% renewable power by 2050. Two scenarios are produced: the Reference Scenario and the Energy Advanced Revolution Scenario (E[R]ev). The E[R]ev looks at the potential of reducing energy demand through energy efficiency and providing electricity through decentralised renewable energy sources. It explores the use of smart grids to connect these decentralised systems and assumes that hybrid electric cars will be predominant in 2050 and that nuclear energy will be phased out by then. Under the E[R]ev, worldwide primary energy demand will decrease to 481.050 petajoules (PJ: energy unit equal to 1.0E+15 joules) in 2050 as a result of energy efficiency measures, the phase-out of nuclear power and a reduced dependence on fossil fuels. Meanwhile, the share of renewable energy will increase considerably, accounting for 82% of worldwide primary energy demand in 2050 (Fig. 3.3). Nearly 95% of the world's electricity is projected to come from renewable energy sources in 2050, reflecting the fact that a renewables-based future is possible (Fig. 3.5). Wind, solar PV and geothermal power are projected to represent approximately 60% of electricity generation and approximately 91% of heat energy is projected to come from renewables. Due in a large part, to the growth of offshore wind energy (to 892 GW in 2050), wind is expected to become the leading renewable energy source, followed by solar PV and solar thermal. Meanwhile, ocean energy will also expand significantly, surpassing geothermal and biomass power in installed capacity. In terms of geography, OECD developed countries are expected to lead in renewable energy production. Four scenarios with significant policy, business and civil society impact due to the high share of renewables in the electricity production were designed by the European Climate Foundation, Greenpeace and the European Commission. The percentage of electricity production from renewables in alternative energy scenarios (including the Energy Advanced Revolution scenario (EREC et al. 2012) and the 2050 Low-Carbon Economy from the European Climate Foundation [ECF 2010]) is projected to increase from 19% in 2012 to between 40% and 90% by 2050 (Fig. 3.6). In fact, the European Union (EU) is committed to reducing greenhouse gas emissions to 80–95% below the 1990 levels by 2050. To meet this climate target, the EU has established aggressive policy targets for the shares of electricity from different sources, such as 60–70% of electricity from renewables by 2050. In addition, all EU countries have individual targets for the RES share of electricity, ranging from 30 to 90%. Most scenarios project large shares (20–50%) in the electricity production profile from solar photovoltaic, wind, and solar thermal power (Fig. 3.7).

Figure 3.6 provides an overview of the share of different power sources in EU future electricity scenarios to 2050.

Figure 3.7 provides an overview of the share of electricity generation in 2050, by renewable energy source, for different EU future electricity scenarios.

Pessimistic assumptions regarding renewable power generation come from the Outlook for Energy: a view to 2040 report (ExxonMobil

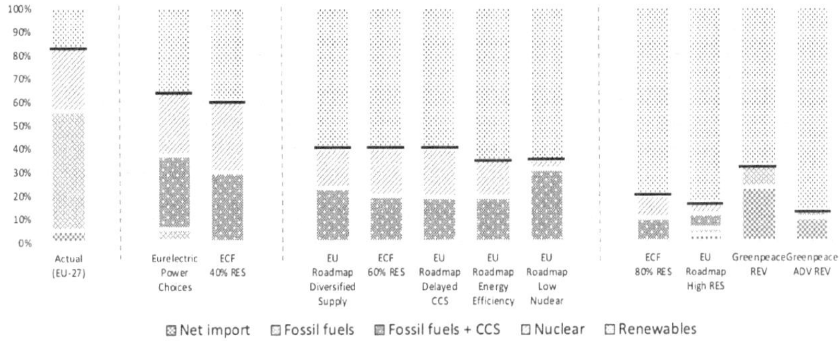

Fig. 3.6 Share of different energy sources, in the generated electricity in 2050, considering future EU electricity scenarios, in % (*Source* Compiled by the authors)

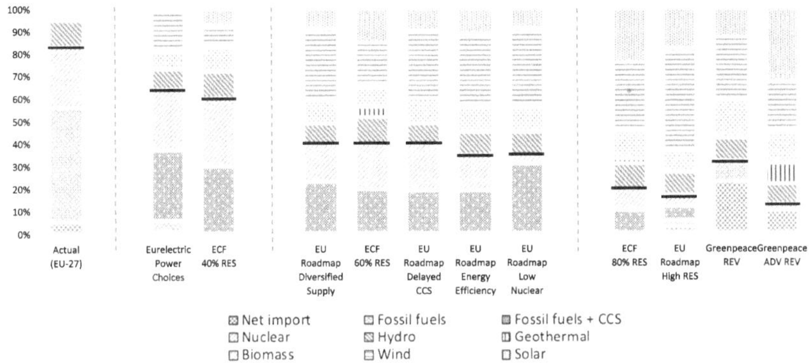

Fig. 3.7 Share of different RES, in the generated electricity in 2050, considering future EU Electricity Scenarios, in % (*Source* Compiled by the authors)

2012); the BP Energy Outlook 2030 report (BP 2013); and the Future World Energy Scenarios (Enerdata 2017). According to ExxonMobil's report, global energy demand will increase by about 30% between 2010 and 2040. Electricity generation will remain the single biggest driver of demand, accounting for more than 40% of global energy consumption by 2040, followed by industry and transport. Energy intensity (calculated as units of energy per unit of GDP) will decrease by almost 45%. The share of fossil fuels in worldwide energy demand will remain very high, at nearly 78%, and the share of RES will increase slowly to nearly

15%; BP is projecting the same trends. Fossil fuels are projected to meet approximately 80% of worldwide primary energy consumption, whilst the share of modern renewables (including biofuels) is projected to be only about 6% in 2030. Renewable energy will grow by about 8% annually between 2010 and 2030. By 2030, RES (excluding hydro) will supply 11% of the world's electricity. According to the Future World Energy Scenarios (Enerdata 2017), the share of fossil fuels in electricity generation is projected, in 2040, to stay at 76% in the Ener-Brown scenario and 71% in the Ener-Blue scenario but will decrease to 50% in the Ener-Green scenario. Meanwhile, RES energy and nuclear energy vary from 24% (Ener-Brown scenario) to 50% (Ener-Green scenario) (Enerdata 2017).

Nuclear Power

Scenarios considered before the nuclear accident at the Fukushima power plant, which caused an earthquake followed by a tsunami, show an increase of nuclear power in the general energy mix. However, after the Fukushima accident, nuclear deployment in the long term, has been revised in many scenarios and national energy policies. For example, Germany announced the immediate closure of eight of its oldest reactors (built before 1980) and the decision to close the rest by 2022, Switzerland has banned the construction of new reactors and Belgium is considering phasing out its nuclear plants. Meanwhile, the Italians have voted overwhelmingly to keep their country non-nuclear.

The WEO (2012) released a scenario with limited development of nuclear power, assuming that no new reactors will be built in OECD countries and that non-OECD countries will only continue with half of their planned nuclear projects. The share of nuclear power is projected to decrease from 13 to 7% in 2035. This decrease is compensated by an increasing use of RES but also fossil fuels. However, it seems difficult to predict the future of nuclear power. Currently, developing nations including Russia, Turkey, Kazakhstan, some European countries (the UK, France, Slovakia, Romania, Poland), Asian countries (China, India, South Korea, Bangladesh, Pakistan) and even Middle East countries (Egypt, Jordan, Iran, UAE) are planning to build new nuclear power facilities.

Transport Sector

In the transport sector, the E[R]ev scenario predicts that electric cars will seriously penetrate the transport sector (Greenpeace 2012), with electric vehicles playing an important role in improving energy efficiency in transport and in replacing fossil fuel vehicles. According to the E[R]ev scenario, in 2030, electricity will provide 12% of the transport sector's total energy demand, whilst in 2050, the share will be 44%, replacing gasoline and diesel. According to Energy Technology Perspective (IEA 2012), the use of electricity in the transport sector will amount to 11% of the overall electricity demand by 2050.

Under the 100% Renewable Energy scenario (WWF and ECOFYS 2011), in the transport sector, there will be a modal shift from fuel to electricity; electric cars and electric rail systems will be more prominent, with more efficient technologies. The basic assumptions for the electrification of transport are a shift to plug-in hybrid and electric cars, energy efficient heavy vehicles, with mainly lithium-ion batteries used for electric vehicle batteries. The electrification of transport will enable this sector to be 100% powered by renewables.

Construction Sector

According to the ETP report, in the Baseline scenario, global final energy demand in the construction sector is projected to increase by 60% between 2007 and 2050 (OECD and IEA 2012), with the total energy demand increasing from 2759 Mtoe (Million Tonnes of Oil Equivalent) in 2007 to 4407 Mtoe in 2050. The residential sector is expected to account for 59% of this growth and the service sector for around 41%. Furthermore, the energy mix in this sector is expected to be dominated by the use of natural gas and, to a lesser extent, by coal. In the BLUE Map scenario, by setting the goal of reducing global energy-related carbon emissions by 50% by 2050 (compared to the 2005 level), energy consumption in the construction sector is expected to decrease by around one-third of the Baseline scenario level by 2050, whilst the consumption of fossil fuels is projected to decline significantly (OECD and IEA 2012). Moreover, solar power is projected to achieve the most growth, accounting for 11% of the total energy consumption in the sector. Its widespread deployment for water heating (30–60% of useful demand depending on the region) and, to a lesser extent, space heating (10–35% of useful demand depending on the region) will assist in improving the efficiency of energy use.

Under the 100% Renewable Energy by 2050 (WWF and ECOFYS 2011) in the construction sector, actions are projected to be taken in the existing building stock and in new buildings. In the existing stock, a retrofitting programme, to be completed by 2050, is key to reducing heating needs by 60%. Heating and hot water needs are expected to be met by solar thermal systems and heat pumps, whilst cooling will be provided by local renewable solutions. New buildings will be near-zero-energy and all-electric by 2030, with residual heat demand met by passive solar, solar thermal installations and heat pumps. Although there will be a decrease in heating demand, a rise in electricity demand will be inevitable as it is expected that buildings will be 'all-electric'. These changes will contribute to buildings being powered 100% by renewables in 2050, compared to a 60% dependence on fossil fuels in 2010.

Carbon Emissions Reduction

Table 3.6 and Fig. 3.8 provide a comparison of various CO_2 mitigation options and their expected contributions in 2050, according to the ETP scenarios (IEA 2012). By far the largest contributor to CO_2 emissions reduction is energy efficiency, accounting for between 31 and 53% of the reduction (below the baseline of 58 Gt by 2050) across all sectors. The ETP scenarios expect Carbon Capture and Storage (CCS) technologies to contribute only between 20 and 28% of the total CO_2 emissions reduction. As far as enhanced-oil-recovery (EOR) is concerned, ETP scenarios expect the potential for EOR to be relatively small, compared to the global emissions from the power generation sector. Fuel switching—defined as the change from traditional solid fuels and modern non-solid fuels—is third (11–16% reduction potential) amongst the CO_2 mitigation options.

Table 3.6 Share of different mitigating actions, in CO_2 reduction, in 2050

CO_2 reduction option	Share in total CO_2 reduction (%)
Energy efficiency	31–53
Carbon capture and storage	20–28
Fuel switching	11–16
Renewable energy in power production	5–16
Nuclear energy	2–10
Biofuels in transport	~6

Source Adapted by the authors from IEA (2012)

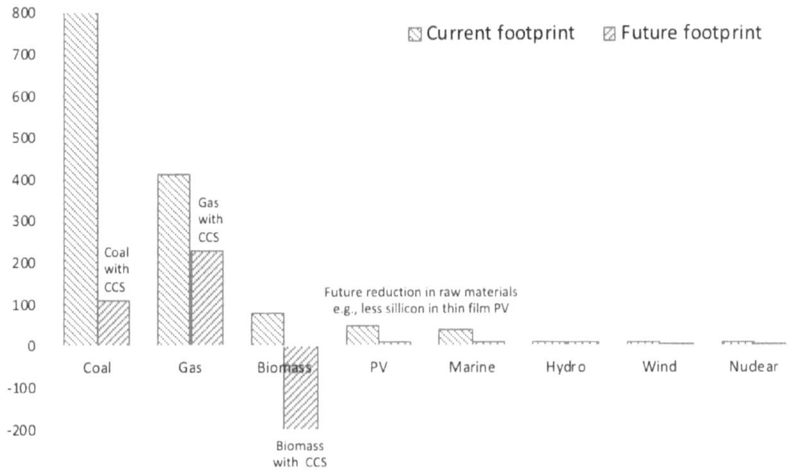

Fig. 3.8 Current and future carbon footprint by energy resource, in Gt (*Source* Adapted by the authors from IEA 2012)

In terms of technology, the carbon emission footprint varies widely within each technology. Overall, nuclear, wind and hydropower have the lowest CO_2e emissions impact (Fig. 3.8), whilst marine power and photovoltaics also have low emissions. For photovoltaics, most of the emissions are the result of electricity use during manufacturing and using raw materials. Coal and gas power have lower CO_2e emissions with CCS technologies. Biomass may also be used increasingly with CCS technologies. CCS technologies have been proposed as a potential method to allow the continued use of fossil-fuelled power stations whilst preventing emissions of CO_2 from reaching the atmosphere and are currently considered technically feasible on a commercial scale, using a range of technologies. CCS involves the CO_2 capture at the point of generation, compression to a supercritical fluid, and sequestration. To reiterate, the CCS methodologies comprise three steps: CO_2 capture/compression/sequestration, CO_2 transportation and CO_2 storage.

Table 3.6 outlines the share (in %) in the total CO_2 reduction for alternative mitigating actions.

Figure 3.8 provides an overview of the current and future carbon footprint of different energy resources including coal, gas, biomass, solar PV, marine, hydro, wind and nuclear.

Transition Costs

The Stern Report (Stern 2007) disseminated to a large audience the conclusions of the last IPCC report relating to emissions reduction costs, incurred mainly by the deployment of low-carbon technologies, including bioenergy with CCS technologies. The Fourth Assessment Report (AR4) showed that on average, the most severe target would cost 3% of worldwide GDP in 2030 and 5–10% in 2050 (Stern 2007). Similarly, to the Stern study, Strachan et al. (2008) find electricity generation a key sector for decarbonisation and highlight the role of critical zero-carbon technologies, including CCS. In general terms, a target of an 80% reduction, compared to the 1990 levels, in 2050 is expected to lead to macroeconomic losses between 0.3 and 1.5% of the world GDP (Strachan et al. 2008).

Similarly, the European Commission's energy project ADAM (Adaptation and Mitigation. Strategies—Supporting European Climate Policy) concluded that losses of GDP, relative to the baseline, vary from about 0.9–2.5% by 2100, compared to a range of 0.5–0.9% in the case of the 550 ppm Stabilisation Scenario, in which a mixture of cap-and-trade systems, sectoral agreements and national measures are adopted (EC 2011). Furthermore, Luderer et al. (2012) show that if the international community takes immediate action to mitigate climate change, the costs of stabilising atmospheric CO_2 concentrations in the case of 450 ppm Stabilisation Scenario will be 1.4% of the world GDP.

They also show that significant investments in the energy sector (of about 9.3 trillion USD to 2030) will be required. A delay in implementing an appropriate climate policy and restrictions to the deployment of low-carbon technologies could result in substantial increases of the mitigation costs.

Discussion

Despite the fact that all scenarios show that RES have the potential to become a major contributor to energy security and sustainable growth in both developed and developing countries, there are considerable uncertainties that could impact on the technical, economic and, in particular, the implementation potential of renewable energy technologies. It is therefore important that a wide range of factors with the potential to cause uncertainty are considered in scenario development and analysis as well as policy design and implementation.

Firstly, in the case of RES deployment, there are uncertainties rising from fuel and carbon emissions trading price fluctuations. In fact, even slight fluctuations in the price of fossil fuels could have wide-reaching impacts on energy and climate policies in developed and developing countries (Kalyuzhnova and Pomfret 2017). For example, environmental programs in resource-rich countries depend on revenues from the oil and gas industry. Secondly, there are social uncertainties. Economic and population growth are two socio-economic factors that affect future energy outcomes, whereas the social acceptability of energy solutions is key to the penetration of novel technologies. For example, there has been social opposition to energy infrastructure for new technologies (such as electricity grids, renewables, nuclear or shale gas) in some EU countries, Japan, South Korea and the United States (Devine-Wright and Batel 2017). Thirdly, there are technological uncertainties (Venetsanos et al. 2002). The price of photovoltaic panels has dropped significantly over the last 20 years and this will continue into the future. However, it is unclear how it will impact on the development of the power sector as a whole and the well-established oil and gas sectors. Furthermore, energy generation and demand, energy efficiency and economic forecasting are other uncertainties that need to be regularly reviewed and analysed (Al-Saleh 2009). Finally, the most challenging uncertainties are those concerned with policy (Barradale 2010). When governments change, their energy policies are also subject to change; reducing policy uncertainty is a crucial component of effectual scenario analysis and effective renewable energy development.

Conclusion

The analysis of medium- and long-term energy scenarios in this chapter shows different visions of a low-carbon energy system in 2050. The scenario assumptions vary significantly regarding the fuel mix, fuel prices, decreases in the cost of technology, future policy actions and transition to low-carbon energy costs. Despite these differences, all energy scenarios considered share some conclusions, mainly that the energy system in approximately 30–50 years will be different from that of today with RES potentially playing a more significant role. Moreover, most of the studies—with the exception of the ExxonMobil report (ExxonMobil 2012) and the BP Energy Outlook report (BP 2013)—agree that there is significant unused potential for renewable energy and that low-carbon energy scenarios, in which additional policies on renewable energy and energy efficiency, as well as carbon cap-and-trade systems are adopted, are plausible ways forward.

All scenarios consider the use of RES an effective way to reduce carbon emissions and to provide significant proportions of the global electricity supply by 2050. Whilst hydropower will maintain its important role, other RES, such as wind, solar and biomass with carbon capture and storage, will grow rapidly in most scenarios. The installation of RES energy facilities in buildings is expected to expand, whilst in the transportation sector many scenarios consider biofuels to be the principal alternative to fossil fuels. Furthermore, the cost of renewable energy technologies is expected to decrease and, within a few years, will most likely make the use of RES more competitive than the use of fossil fuels. This will directly affect the cost of electricity generation and, as a consequence, the profitability of fossil fuels' extraction, influencing the development of the power sector as a whole and the entrenched oil and gas sectors.

The key requirement in low-carbon/high-renewables energy scenarios is extensive government involvement in the promotion of renewable energy technologies, whilst the principal benefit is enhanced energy security, due to reduced carbon emissions, the growing diversification of the energy mix and consequently, the diminished reliance on fossil fuels. However, future policies for strengthening energy security should not focus on increasing the utilisation of RES alone. Rather, they should seek to complement initiatives and instruments for enhancing the penetration of RES by targeting improved energy efficiency and efficient trading of carbon emissions. It is by addressing the multidimensional nature of energy security within the relevant context, be it local, regional or national, that energy security can be bolstered in a sustainable fashion.

Acknowledgements This research was supported by the European Commission project MILESECURE-2050 (Multidimensional Impact of the Low-carbon European Strategy on Energy Security, and Socio-Economic Dimension up to 2050 perspective), with additional support from Kazakhstan's Ministry of Science' grant AP05135081 (Energy Security and Energy Policymaking Framework in the Eurasian Economic Union).

References

Al-Saleh, Y. (2009). Renewable energy scenarios for major oil-producing nations: The case of Saudi Arabia. *Futures, 41*(9), 650–662.

Ang, B. W., Choong, W. L., & Ng, T. S. (2015). Energy security: Definitions, dimensions and indexes. *Renewable and Sustainable Energy Reviews, 42,* 1077–1093.

Barradale, M. J. (2010). Impact of public policy uncertainty on renewable energy investment: Wind power and the production tax credit. *Energy Policy, 38*(12), 7698–7709.

Bassam, N. (2001). Renewable energy for rural communities. *Renewable Energy, 24*(3), 401–408.

Bloomberg. (2015, June). *New energy outlook 2015—Long-term projections for the global energy sector.* Executive Summary. Available at: https://www.seia.org/sites/default/files/resources/BNEF-NEO2015_Executive-summary.pdf.

BP. (2013). *BP energy outlook 2030.* Available at: https://www.bp.com/content.

Devine-Wright, P., & Batel, S. (2017). My neighbourhood, my country or my planet? The influence of multiple place attachments and climate change concern on social acceptance of energy infrastructure. *Global Environmental Change, 47,* 110–120.

European Climate Foundation (ECF). (2010). *Roadmap 2050: A practical guide to a prosperous, low-carbon Europe.* European Climate Foundation. Available at: http://www.roadmap2050.eu/attachments/files/Volume1_full-report_PressPack.pdf.

Ellabban, O., Abu-Rub, H., & Blaabjerg, F. (2014). Renewable energy resources: Current status, future prospects and their enabling technology. *Renewable and Sustainable Energy Reviews, 39,* 748–764.

Enerdata. (2017). *Future world energy scenarios.* Available at: https://www.enerdata.net/research/forecast-enerfuture.html.

European Commission (EC). (2011). *ADAM: ADaptation And Mitigation strategies supporting European climate policy.* Available at: http://www.tyndall.ac.uk/adamproject.

European Renewable Energy Council (EREC), Global Wind Energy Council (GWEC), & Greenpeace International. (2012). *Energy revolution: A Sustainable world energy outlook 2012.* Available at: http://www.greenpeace.org/.

ExxonMobil Corporation. (2012). *The outlook for energy: A view to 2040.* Available at: https://www.exxonmobil.com.

Francés, G. E., Marín-Quemada, J. M., & González, E. S. M. (2013). RES and risk: Renewable energy's contribution to energy security. A portfolio-based approach. *Renewable and Sustainable Energy Reviews, 26,* 549–559.

Global Wind Energy Council. (2013). *Global wind energy outlook.* Available at: http://www.gwec.net/publications/.

International Energy Agency (IEA). (2012). *World energy outlook 2012.* International Energy Agency. Available at: https://www.iea.org/publications/freepublications/publication/English.pdf.

International Institute for Applied Systems Analysis (IIASA). (2012). *Global energy assessment: Toward a sustainable future 2012.* Available at: http://pure.iiasa.ac.at/id/eprint/10099/.

IPCC. (2007). *Climate change 2007: Synthesis report. Contribution of working groups I, II, and III to the fourth assessment report of the intergovernmental panel on climate change.* Cambridge: Cambridge University Press.

Kalyuzhnova, Y., & Pomfret, R. (2017). Resource-rich countries, clean energy and volatility of oil prices. In Y. Kalyuzhnova (Ed.), *Sustainable energy in Kazakhstan* (pp. 25–35). Routledge.

Lombardi, P., & Gruenig, M. (2016). *Low-carbon energy security from a European perspective*. Cambridge, MA: Academic Press.
Luderer, G., Bosetti, V., Jakob, M., Leimbach, M., Steckel, J. C., Waisman, H., et al. (2012). The economics of decarbonizing the energy system—Results and insights from the RECIPE model intercomparison. *Climatic Change, 114,* 1–29.
Martinot, E., Dienst, C., Weiliang, L., & Qimin, C. (2007). Renewable energy futures: Targets, scenarios, and pathways. *Annual Review of Environment and Resources, 32,* 205–239.
National Renewable Energy Laboratory (NREL). (2015). *NREL System Advisor Model (SAM)*. Available at: https://sam.nrel.gov.
Ölz, S., Sims, R., & Kirchner, N. (2007). *Contribution of renewables to energy security*. Paris: International Energy Agency, OECD/IEA.
Organisation for Economic Cooperation and Development and International Energy Agency (OECD and IEA). (2012). *Energy technology perspectives: Pathways to a clean energy system* (2012th ed.). Paris: OECD and IEA. Available at: https://www.iea.org/publications/freepublications/publication/ETP2012_free.pdf.
Panwar, N. L., Kaushik, S. C., & Kothari, S. (2011). Role of renewable energy sources in environmental protection: A review. *Renewable and Sustainable Energy Reviews, 15*(3), 1513–1524.
Söderholm, P., Hildingsson, R., Johansson, B., Khan, J., & Wilhelmsson, F. (2011). Governing the transition to low-carbon futures: A critical survey of energy scenarios for 2050. *Futures, 43*(10), 1105–1116.
Stern, N. (2007). *The economics of climate change: The Stern review*. Cambridge: Cambridge University Press.
Strachan, N., Foxon, T., & Fujino, J. (2008). *Modelling long-term scenarios for low carbon societies*. London: Earthscan; James and James.
Umbach, F. (2010). Global energy security and the implications for the EU. *Energy Policy, 38*(3), 1229–1240.
United Nations (UN). (2014). *World urbanization prospects: The 2014 revision. Highlights* (p. 32). New York: United Nations. Available at: https://esa.un.org/unpd/wup/publications/files/wup2014-highlights.pdf.
United Nations (UN). (2015). *The Paris Agreement on climate change* (p. 27). New York: United Nations. Available at: https://unfccc.int/sites/default/files/english_paris_agreement.pdf.

United National Development Programme (UNDP). (2010). *World population prospects: The 2010 revision. Volume I: Comprehensive Tables* (p. 482). New York: United Nations, 2011.
United Nations Framework Convention on Climate Change (UNFCCC). (2011, February). *Fact sheet: The Kyoto Protocol*. New York: United Nations. Available at: https://unfccc.int/files/press/backgrounders/application/pdf/fact_sheet_the_kyoto_protocol.pdf.
U.S. Energy Information Administration (US EIA). (2011). *International energy outlook 2011*. Washington, DC: US EIA, DOE/EIA-0484(2011)/September 2011. Available at: https://www.eia.gov/outlooks/archive/ieo11/pdf/0484(2011).pdf.
Venetsanos, K., Angelopoulou, P., & Tsoutsos, T. (2002). Renewable energy sources project appraisal under uncertainty: The case of wind energy exploitation within a changing energy market environment. *Energy Policy, 30*(4), 293–307.
World Wild Fund (WWF), & ECOFYS. (2011). *The energy report: 100% renewable energy by 2050*. Available at: http://www.ecofys.com/.

4

Enabling Green Energy Production: Implementing Policy by Using Public–Private Collaboration

Nikolai Mouraviev and Anastasia Koulouri

Introduction

Since 1991, when Kazakhstan became an independent nation, its economic development has been driven by revenue raised from the sale of natural resources, principally oil, coal and uranium. In various years since the beginning of the 1990s, revenue from the sale of natural resources has formed between 50 and 75% of the government budget, and oil extraction has continuously expanded over time. Much like in most resource-rich nations, Kazakhstan's dependency on oil revenue and the world oil price has created incentives to extract and sell ever-increasing volumes of oil. Despite a few government programmes launched since 2000 aimed at restructuring the economy, no new industries have been created in recent years. Furthermore, in the past ten years,

N. Mouraviev (✉) · A. Koulouri
Dundee Business School, Abertay University, Dundee, UK
e-mail: n.mouraviev@abertay.ac.uk

A. Koulouri
e-mail: a.koulouri@abertay.ac.uk

N. Mouraviev and A. Koulouri (eds.), *Energy Security*,
https://doi.org/10.1007/978-3-030-01033-1_4

the world oil market has presented new challenges to all oil-exporting nations including Kazakhstan: rapid changes in the price of oil jeopardise planned formation of the government budget and, therefore, put all public sector funding at risk. During the 2007–2009 economic crisis and subsequent macroeconomic fluctuations (including the recent economic downturn that began in 2014), oil revenue has been shrinking due to the rapidly falling oil price that was, in 2016, at times as low as USD 25 per barrel.

In the light of unstable oil prices, which make the revenue side of the government budget highly uncertain, Kazakhstan has been actively seeking ways to attract private investment that might complement and/or substitute government spending on public services. Beginning in 2006, the government adopted public–private partnerships (PPPs) as a tool that can bypass budgetary limitations, create jobs and deliver public services in many industries including the energy sector (Mouraviev et al. 2012). A PPP exists where a private investor builds or renovates an asset at its own expense and subsequently provides a service with the use of this asset over an extended period, normally between 15 and 30 years (Mouraviev and Kakabadse 2016). In order to recover its investment and make a profit, a private operator collects user fees and/or payments from the government. This chapter discusses whether Kazakhstan's current extensive oil extraction, and corresponding reliance of the government budget on oil revenue, can be reduced via the use of PPPs in the energy sector.

Aim, Scope and Structure

Certain countries including Germany, the Netherlands, Portugal and Spain have used PPPs in order to promote the production of energy from renewable energy sources (RES) (Reiche and Berchberge 2004). In this chapter, the term 'promotion of RES' includes the expansion of both the production and the use of power from renewable sources. Furthermore, 'green energy' (or 'clean energy') is the same as power from RES, and a 'green economy' is an economy that increasingly relies on RES, rather than fossil fuels. In summary, the chapter's scope includes the production and consumption of green energy and all activities related to bringing green energy to consumers, while promotion refers to all forms of government actions and initiatives intended to expand the use of renewable sources for energy generation.

Drawing on the experience of these and other nations, this chapter aims to investigate what role PPPs can play in the promotion of RES, for example by building wind energy plants and municipal solid waste management (waste-to-energy) facilities. Specifically, the objective is to consider whether PPPs, as a policy instrument, could be effectively used in Kazakhstan for the sustainable production of power from RES. In the context of this chapter, 'sustainable use of RES' means the purposeful and ever-increasing utilisation of renewable sources for the production of power, and a corresponding decrease in reliance on non-renewable sources, with a long-term strategic goal to ensure the continuous expansion of the use of RES.

This chapter begins by surveying the experience of selected economies in the promotion of green energy and puts forward critical questions regarding the role that PPPs might play in the sustainable production of power from renewables. The next section gives details of PPP development in Kazakhstan from 2005 to present. Then Kazakhstan's efforts in promoting green energy are highlighted, which includes an analysis of the nation's energy mix, delineation of the legal framework for RES, and comments regarding how the policy on green energy is implemented. This shows the significance of expanding the RES utilisation in the context of the nation's economy and elucidates Kazakhstan's preparedness for this expansion. The next section identifies the conditions for successful PPP implementation. The rationale is that PPPs will be able to serve as an effective policy tool for the promotion of green energy when certain conditions are in place. As current PPPs in Kazakhstan operate within a certain legal and regulatory framework, investigating how well these institutionalised elements meet the objective of promoting green energy becomes the next step in analysing PPPs' suitability for the sustainable production of green energy. The fit between the existing legal and regulatory PPP framework that forms the platform for successful PPP implementation, on the one hand, and an emergent policy on green energy, on the other, is then assessed using interview data collected from actors in the government and ongoing

PPP projects in Kazakhstan. To summarise, the chapter makes use of earlier research conducted by the authors on PPP management in Kazakhstan and draws on these data through the lens of potential PPP suitability for the sustainable production of power from RES.

PPPs as a Policy Tool for the Promotion of Green Energy: The Experience of Selected Nations

This section highlights the experience of certain economies that have achieved progress in using RES. An assessment of these experiences will help to understand the role that PPPs might play in the production of energy from renewable sources.

The experience of nations, such as Brazil, Canada, Germany and the UK, shows that a required precondition for promoting RES is the government's political decision regarding the use of green energy (Wachsmann and Tolmasquim 2003; Liming et al. 2008; Valle Costa et al. 2008). However, deciding on a direction for the energy policy is not enough: the general direction should be complemented by clear goals, and the government should set specific targets. For example, the government may require an annual increase in the production of green energy as a percentage of the total electricity production, i.e. that year-by-year the share of power from RES should be increasing in the total volume of generated energy (Stenzel and Frenzel 2008). Once these two fundamental elements—policy direction and specific targets—are in place, this effectively forms public policy that fosters the diversification of energy sources and facilitates the employment of a range of policy instruments for the ever-increasing production of green energy (Menz and Vachon 2006; Dinica 2011).

Among the commonly used instruments for increasing the use of RES are price support systems, subsidy schemes, low-interest loans and fiscal incentives (e.g. tax exemption for all domestic generators of renewable energy) (Ackermann et al. 2001; Dutra and Szklo 2008). Yet another commonly employed tool is the feed-in tariffs that offer a green energy producer a guaranteed payment of a certain percentage of a price that a consumer pays and permit the feed of a producer's renewable energy into a nation's electrical power grid (Agnolucci 2006).

In addition, governments often use another policy tool: they deploy PPPs for the purpose of power generation and transmission. In the energy sector, PPPs may be deployed in order to use non-renewable sources (i.e. stations that use oil and gas for energy generation) as well as facilities that use RES (e.g. stations that use wind or solar energy for the production of power). An analysis of PPPs' effectiveness in green energy production in Spain (Dinica 2008), Portugal (Martins et al. 2011) and France (Bougrain 2012) raises the argument that, although PPPs are viewed as a powerful tool to promote RES, partnerships have to be underpinned by public policy and a variety of legislative norms and regulatory arrangements. For example, in Portugal, the government emphasised that the main criterion for a successful PPP bid is the creation of an industrial cluster, rather than the production of electricity from RES (Dinica 2008). It is worth noting that putting together a successful bid, although ensuring compliance with the government's requirements, does not necessarily guarantee the successful delivery of what the government aims to achieve in terms of the expanded use of RES. In other words, the assessment criteria for PPP bids may or may not relate to achieving effective use of PPPs for the production of green energy. Therefore, PPPs should be considered not just an instrument for the promotion of the expanded use of RES, but rather a policy tool that is part of a government's broader framework for diversification of the economy and sustainable development (i.e. PPPs may be deployed for a variety of purposes, and production of power from RES is just one of them). To reiterate, the multiple purposes for which PPPs might be launched should be integrated with the government's broader long-term strategy for sustainable development.

Considering the PPPs' role in relation to RES, the following questions should be addressed:

- Could PPPs contribute to the implementation of Kazakhstan's policy aiming to promote the production and use of green energy? If so, under what conditions?
- Do PPPs fit the government's broader goals of diversifying the economy and reducing the nation's dependency on oil and oil revenue?

PPPs in Kazakhstan

From 2005 to the present, the Kazakhstan's government has institutionalised PPPs as a policy tool for the delivery of public services in social and industrial infrastructure. PPP development in Kazakhstan began in 2005–2006, when in 2006 the government approved the law on concessions (Zakon Respubliki Kazakhstan 2006). Subsequently, the government created a PPP institutional framework by establishing, in 2007, the National PPP Centre, followed by the formation of PPP centres in selected regions in Kazakhstan. The National PPP Centre, being formed by and attached to the Ministry of Economic Development, has played a key role in preparing a large number of projects for which the government seeks investors. The Centre has also drafted standard PPP contracts that could be used in a variety of sectors (e.g. toll roads or kindergartens) and aimed to ease and accelerate PPP implementation. The proposed PPP projects and standard contracts are available on the Centre's website (Kazakhstan Public–Private Partnership Centre 2016). Overall, the Centre has been an influential facilitator of the PPP development in Kazakhstan.

Despite active facilitation by the National PPP Centre, progress towards PPP deployment has stalled: as of January 2017, there are only three ongoing PPP projects in Kazakhstan. In the transportation sector, these are a 120 km segment of the railroad in Eastern Kazakhstan between Shar and Ust-Kamenogorsk and the construction and operation of a passenger terminal in an international airport in Aktau, a city on the Caspian Sea coast. In the energy sector, the concession involves the construction and operation of an interregional electrical grid between Northern Kazakhstan and the Aktobe region.

In an attempt to improve the PPP legislative framework and give new impetus for public-private collaboration, in 2015 the Kazakhstan's government passed the new Law on Public–Private Partnerships, which came into effect in 2016. The law specifies two forms for a PPP: contractual and institutional. The latter is understood as a firm that is formed jointly by the public and private sector partners as a joint stock company (i.e. a corporation) or a limited liability partnership. In addition to a concession, the law provides for the possibility of other PPP forms (Mouraviev and Kakabadse 2017). The law made new partnership forms available for a PPP launch, and the RES sector is likely to be one of the industries that will draw significant government support and private investment in the future.

Kazakhstan's Emergent Policy on Green Energy

This section includes two parts: the first part outlines the country's energy profile demonstrating its energy mix and its dependency on fossil fuels; and the second part provides an overview of the Kazakhstani government's efforts to promote green energy through policy initiatives, the creation of an appropriate legal framework, and other developmental steps.

Kazakhstan's Energy Profile

As Kazakhstan is rich in fossil fuels (oil, gas and coal), uranium for nuclear power, as well as in renewable energy sources, it is among the largest energy producers in Central Asia (International Energy Agency 2015).

In 2012, Kazakhstan ranked 12th in the world for crude oil reserves, 18th for natural gas reserves and 8th for hard coal reserves (BGR 2013). It is among the top eight lignite producers worldwide (BGR 2013) and has the third-largest deposits of uranium, extracting a third of the world's total uranium output, which makes the nation the top producer worldwide. At the same time, Kazakhstan shows significant potential for using RES (including solar, wind, biomass, hydro and geothermal), which is estimated at over one trillion kWh per year, with wind and solar resources that are a viable alternative for power generation both technically and economically (Energy Charter Secretariat 2013; REEEP 2014).

The country's energy production, Total Primary Energy Supply[1] (TPES) and Total Final Consumption[2] (TFC) are dominated by fossil fuels with 50.72 and 30.03% of the energy production having been contributed by oil and coal, respectively; 48.31 and 33.85% of the TPES having been contributed by coal and natural gas, respectively; and 29.02 and 28.42% of the TFC having been attributed to coal and oil products/oil, respectively (International Energy Agency 2016). Less than 0.5% of Kazakhstan's energy production mix can be attributed to RES, 90% of which is provided by small hydropower plants (Energy Charter Secretariat 2013). Moreover, less than 1% of the TPES mix and less than 0.1% of the TFC mix are contributed by RES (TPES: 0.43% hydro, negligible contribution from other sources; TFC: 0.06% biofuels) (International Energy Agency 2016). In 2014, 7.86% of the country's electricity generation mix was provided by hydropower and wind power has just started contributing to the mix (International Energy Agency 2015, 2016). Nevertheless, Kazakhstan has the second-lowest contribution of RES in the TPES mix among the countries of Eastern Europe, Caucasus and Central Asia (EECCA is a block of countries that includes Armenia, Azerbaijan, Belarus, Georgia, Kazakhstan, Kyrgyzstan, Moldova, the Russian Federation, Tajikistan, Turkmenistan, Ukraine and Uzbekistan) (International Energy Agency 2014). Table 4.1 shows the percentage contribution of different elements to Kazakhstan's energy production, TPES, TFC and electricity generation.

Although Kazakhstan is a resource-rich country, there are growing concerns regarding its overreliance on fossil fuels and corresponding

---

[1]TPES: Total Primary Energy Supply is made up of the indigenous production plus imports minus exports, international marine and aviation bunkers, and plus/minus stock changes (International Energy Agency, https://www.iea.org/statistics/resources/balancedefinitions/#tpes, accessed 5 Dec 2016).

[2]TFC: Total Final Consumption is "the sum of consumption by the different end-use sectors. TFC is broken down into energy demand in the following sectors: industry, transport, buildings (including residential and services) and other (including agriculture and non-energy use). It excludes international marine and aviation bunkers, except at world level where it is included in the transport sector" (International Energy Agency, https://www.iea.org/about/glossary/t/, accessed 5 Dec 2016).

Table 4.1 Kazakhstan's key energy data[a] (2014)

	Oil (%)	Coal (%)	Natural gas (%)	Hydro (%)	Other
Energy production	50.72	30.03	18.8	0.43	Geothermal, solar, wind, biofuel and waste: negligible
TPES[b]	24.45	48.31	33.8	0.43	Geothermal, solar, wind, biofuel and waste: negligible
TFC	0.90	29.02	7.71	N/A	Oil products: 27.52%; heat: 18.59%; electricity: 16.19%; biofuels and waste: 0.06%
Electricity generation	0.94	71.95	19.20	7.86	

Source Compiled by the authors using data from Kazakhstan: Balances for 2014 (International Energy Agency 2016)
[a]Totals of % may not add up due to rounding
[b]Total also takes into account: −7.44% for oil products and −0.13% for electricity

dependency on the world oil price, which underlies its increasing attention to the utilisation of renewable sources. The next section highlights the government's initiatives to promote green energy.

Kazakhstan's Policy and Legal Framework for the Promotion of Green Energy

- Policy

The European Bank for Reconstruction and Development in its report on "Securing sustainable energy in transition economies" (EBRD 2008) highlighted the need to integrate policies promoting energy efficiency, developing renewable energy sources and mitigating against environmental impact to support long-term economic development and enhance energy security. This demonstrates the international trend that applies not only to Kazakhstan, but also many other transitional as well as industrialised nations.

Kazakhstan's government has recognised the need for a shift from its dependency on fossil fuels to a more sustainable economic model by adopting in 2009 the Law of the Republic of Kazakhstan on Support of the Use of Renewable Energy Sources (2009). In 2012, the President of Kazakhstan, in the strategy Kazakhstan 2050, outlined the social, economic and political reforms that Kazakhstan needs to undertake in order to join the top 30 world economies by 2050, and achieve economic growth by improving the investment climate, diversifying production of goods for export, and supporting the private sector and public–private partnerships. Identifying energy security as one of the top ten global challenges in the twenty-first century, Kazakhstan's President set out the goal that, by 2050, at least 50% of the country's total energy consumption should be covered by alternative and renewable energy sources (Kazakhstan 2050, 2012).

The Kazakhstan 2050 strategy created a new impetus for the government that has subsequently adopted policies and introduced laws to support its implementation. Table 4.2 summarises the policy actions of the Kazakhstani government and the steps it has taken to promote RES in the country.

The Kazakhstan's government's efforts to promote RES began in 2010. They include numerous policy actions and initiatives providing direction and setting specific targets. This effectively has ensured the emergence of a public policy that aims to diversify the use of the country's energy sources and foster its sustainable economic development.

- Legal Framework

The International Energy Agency confirms that Kazakhstan has a "well-elaborated legislation in the field of renewable energy" (International Energy Agency 2015, 172). In addition to a number of

Table 4.2 Green energy in Kazakhstan: key government initiatives and developmental efforts

Year	Initiative/effort	Comments
2010	State programme of electricity sector development for 2010–2014 (Energy Charter Secretariat 2013)	Takes RES into consideration
2010	State programme for accelerated industrial and innovative development for 2010–2014 (2010)	Takes RES into consideration (www.akorda.kz)
2011	Green bridge partnership programme	Regional initiative pioneered by Kazakhstan to facilitate the greening of the economy (https://sustainabledevelopmentsustainabledevelopment.un.org/partnership/?p=2237)
2012	Kazakhstan 2050	Outlines reforms necessary for the economy to join the list of top 30 nations in the world. Identifies energy security as key challenge. Sets the goal of using RES for at least 50% of the total energy consumption, by 2050
2013	Action plan for the development of alternative and renewable energy for 2013–2020	Outlines 31 RES projects (wind farms, solar installations, hydropower plants) with a total capacity of 1040 MW, to be implemented by 2020 (www.globalmethane.org)
2013	Green economy concept	Outlines the nation's new development pattern, the green path of development. Sets the target to increase the share of RES in the energy mix: by 2020, RES should contribute no less than 3% of the total power generation and 50% by 2050 These targets should be achieved by promoting green energy generation through: • Creating favourable conditions for constructing/operating RES facilities; • Integrating RES capacities to the Unified Power System; • Incentivising investment in the sector

(continued)

Table 4.2 (continued)

Year	Initiative/effort	Comments
2013	Green Academy	Research and education centre established by the Ministry of Energy to support the implementation of the Green Economy Concept (http://www.green-academy.kz/en/)
2013	National Programme of Wind Power Sector Development (Energy Charter Secretariat 2013)	Makes wind energy power generation a key priority. Aims to generate up to 5 billion kWh from wind, by 2024. Evolved from the Kazakhstan–Wind Power Market Development Initiative project (2004–2011) that was financed by the Global Environment Facility and implemented by the United Nations Development Programme and Kazakhstan's government (UNDP 2011)
2015	100 Concrete Steps for the Implementation of Five Institutional Reforms	Outlines 100 steps for the implementation of the Kazakhstan 2050 strategy including actions for: • Restructuring the electric power industry; • Expanding the regional electrical grids/networks; • Implementing new electricity tariffs; and • Attracting investors to the industry that makes energy-saving equipment and technology (www.kazembassy.org.uk)
2016	Green Bridge Institute	Created to provide scientific and technological support to green economy projects and serve as a cooperation platform for the Ministry of Energy and the Kazakh National Research Technical University (Government of the Republic of Kazakhstan 2016)
2017	EXPO2017 Future Energy	International exposition hosted in Astana, Kazakhstan, in the summer of 2017. Aims to raise awareness on sustainable development and showcase the latest green energy technologies (https://expo2017astana.com/)

Source Compiled by the authors

legislative provisions governing the power sector,[3] the key legislative act for RES is the Law of the Republic of Kazakhstan on Support of the Use of Renewable Energy Sources (2009). The law was initially introduced in 2009 and then subsequently amended in 2013. It sets the foundation for the government's regulation of the RES sector and provides support mechanisms for the increasing use of renewable sources by stipulating the following:

- New RES energy-generating facilities must be connected to the grid. Regional power grid companies are responsible for connecting green energy-generating facilities to the grid and purchasing all electricity produced from RES;
- As of January 2014, the government guarantees RES power plants feed-in tariffs, including wind (22.68 KZT/kWh), solar with Kazakh silicon (34.61 KZT/kWh), solar with Kazakh modules (70 KZT/kWh), hydropower (16.71 KTZ/kWh), and biogas (32.23 KZT/kWh). These are guaranteed for a 15-year period and will be adjusted by the annual inflation rate;
- The government grants RES energy-generating companies an exemption from payment for power transmission services.

In addition to the provisions of the 2009 law, the country's Land Code (2003) permits regional authorities to allocate land plots for the construction of facilities that aim to use RES. Furthermore, when the government provides land plots for this purpose, investors should be invited to participate in the design and construction of RES facilities (Energy Charter Secretariat 2013).

---

[3] These include: a presidential decree in 1995 establishing the current electricity market structure; the Programme of Privation and Restructuring of the Power Sector (Government Resolution No. 663, 1996); the Law of the Republic of Kazakhstan (RK) about Energy Saving (25 Dec 1997, No. 210-I); the Law of the RK about Electricity (9 July 2004, No. 588-II); the Law of the RK about Gas and Gas Supply (9 Jan 2012, No. 532-IV); the Law of the RK about Energy Saving and Increasing Energy Efficiency (13 Jan 2012 No. 541-IV). Also, the Law of the RK about Subsoil Resources and Subsoil Use (24 June 2010, No. 291-IV); the Law of the RK about Architectural, Town Planning and Construction in the RK (No. 242-II, 2001).

As is the case of electricity and heat generation using fossil fuels, the generation and distribution of electricity and heat from RES should be licensed, and the operation of RES power plants requires authorisation by the Agency of the Republic of Kazakhstan for the Regulation of Natural Monopolies (Energy Charter Secretariat 2013). Potential investors in green energy generation and/or distribution projects might view these requirements as an additional layer of bureaucracy. This perception could become a barrier to attracting investors and impede the progress of the promotion of RES in the country. Furthermore, the International Energy Agency reports that "licencing remains problematic, particularly for small and medium-sized enterprises" (International Energy Agency 2015, 200).

Although in recent years Kazakhstan has achieved progress in the design of the legal framework for renewable energy, it is still relatively new in comparison with the framework relating to energy from fossil fuels. This means that the framework for RES is yet to be tested and its effectiveness will be determined over a number of years. Furthermore, a multiplicity of legal provisions and institutions responsible for their implementation raise concerns regarding the lack of clarity, challenging inter- and intra-agency communication and duplication of effort. This legal and institutional fragmentation could create another barrier in the promotion of RES in Kazakhstan.

- Implementation

To implement the State Programme for Accelerated Industrial and Innovative Development for 2010–2014 (2010), the National Atomic Company Kazatomprom established independent subsidiaries in order to undertake projects in the RES sector. These include: KAZ PV Project (solar), which is constituted by Kvarts LLP (quartz mining), KazSilicon LLP (quartz processing and silicon production), Kaz Solar Silicon LLP (silicon processing and photovoltaic panel production) and AsSolar LLP (photovoltaic module production); Ecoenergomash LLP (wind); Mashzavod LLP and Legmash LLP (heat pumps) (Energy Charter Secretariat 2013).

In addition, in 2012 a government-owned corporation called Samruk-Energo (Samruk-Energy Corporation), a holding company that manages energy enterprises in Kazakhstan, formed a company Samruk-Green Energy (www.samruk-green.kz) to operate in the field of power production from RES. The company aims to:

- Create technology, design and construct facilities that will produce energy and heat from solar power;
- Produce and sell energy and heat from solar power;
- Maintain transmission networks transporting electricity from the production facilities to the distribution networks; and
- Undertake research and development activities and provide consulting services in the field of solar energy.

When compared with the oil and gas sector, that has been responsible for the rapid economic growth that the country experienced in the 1990s and early 2000s, the gradual shift of the government's focus from oil and gas to renewable sources may be viewed as a drain on resources. Although Kazakhstan has made certain steps to form a policy and establish a legal framework in order to promote green energy, progress remains slow and the development of the sector is still in its infancy. The RES sector needs investment, incentives and instruments to ensure technological development, faster deployment of energy-generating facilities and the construction of connecting power networks. One apparent problem in the promotion of RES is the lack of incentives and practical mechanisms that would give impetus to the acceleration of green energy generation. While the policy is in place, implementation tools appear to be lacking.

PPPs as a Policy Tool: Conditions for Effective Use

This section discusses the factors that ensure the effective use of PPPs as a policy tool. From the perspective of stakeholder management, a PPP is a set of complex inter-organisational arrangements, with multiple stakeholders, rather than a stand-alone organisation (Mouraviev and Kakabadse 2017). Therefore, focusing on capturing, trading and serving the values of PPP stakeholders provides an appropriate framework for conceptualising the prerequisites for PPP implementation. At the core of successful PPP implementations are those activities and practices that ensure the effective management of stakeholder values (Yang et al. 2009). The stakeholder management perspective is useful as it accurately highlights societal preferences for PPP deployment. These societal preferences need to be aligned with each other, and include:

- A government's willingness to extend financial and administrative support to a private partner;
- The private investors' interests in long-term profitable projects supported by government guarantees and subsidies;
- Priority sectors for PPP deployment determined by the government;
- Regional needs and preferences; and
- The citizens' needs and willingness to receive PPP-provided public services, as taxpayers should be willing and able to pay often higher tariffs and/or possibly higher taxes.

To determine the conditions for successful PPP implementation, this chapter draws on data from interviews that were conducted in 2012–2016 with a range of PPP actors from both the government and the private sector in Kazakhstan. While the range of conditions might be very broad, the interviewees focused mostly on the legal and regulatory environment for PPPs, which forms prerequisites for PPP deployment. In the opinion of interviewees, this factor grouping includes four components:

- a well-designed legal framework;
- simplified processes and procedures related to PPP formation;
- market discipline (i.e. partners must fulfil their obligations); and
- a clear tariff policy.

A well-designed, detailed and clear legal framework is often noted as a prerequisite for all types of PPP activity (Pongsiri 2002). In interviews conducted prior to 2015, respondents frequently noted the absence of a general PPP law in Kazakhstan and the drawbacks of the 2006 Kazakhstan's Law on Concessions (Zakon Respubliki Kazakhstan 2006). From the interviewees' perspective, this made a PPP an ambiguous form of public–private collaboration. In 2015, the government adopted a new law on PPPs that came into effect in 2016 (Zakon Respubliki Kazakhstan 2015). Although the 2015 law incorporated a very broad definition of a PPP, the law marked a significant step forward by bringing more clarity to the PPP legislative framework. One specific area that, prior to 2015, required legislative clarification was the definition of non-concessionary PPP models, in addition to a concession, which the 2015 law addressed.

Simplifying the PPP formation procedures might significantly intensify PPP development in Kazakhstan. Creating a streamlined single-tender bidding process that facilitates granting a single PPP contract to the winner, rather than a set of different contracts with their own provisions and regulations, such as a contract for land use, another contract for the service provision and a separate partnership contract, could accomplish this. Interviewees expressed criticism by highlighting excessive bureaucracy and tangled procedures in the PPP formation process, which the government needs to simplify and shorten.

Ensuring market discipline is also a critical area for PPP success (Jamali 2004). Market discipline requires either partner, whether from the public or the private sector, to honour payments, work schedules and other obligations as specified by contracts and regulations. For example, either partner must make payments to the other, subcontractors, workers and customers fully, on time and according to contract terms and schedules. Tools such as interest payments, fees and fines applied to either partner should ensure market discipline. Interviewees noted that it would be useful to incorporate these tools in the PPP regulations and define the terms and conditions under which a certain tool can or should be used.

The PPP regulatory environment may become more transparent and attractive to investors once they know and understand a tariff policy for a partnership's services. A tariff policy should include at least three components: conditions for setting a new tariff; regulations regarding ranges of future tariff adjustments; and a clear, streamlined application process for a new tariff. Although interviewees reported that procedures for tariff adjustments are bureaucratic, tangled and lengthy, they can be improved, which will permit private operators to conduct effective long-term business planning. Improvements can be achieved by incorporating a certain section in the regulatory framework that would define the scope, conditions and procedures for tariff adjustments. Subsequently, these provisions could be included, fully or in part, in a contract for a PPP project.

Summarising the interviewees' opinion regarding factors that ensure the successful use of PPPs as a policy tool, the following question needs to be answered: does the legal and regulatory PPP framework in Kazakhstan support the effective deployment of partnerships for the promotion of green energy in the nation? Interviewees noted that certain critical elements of the PPP legal framework are in place and emphasised the significance of the 2015 law on PPPs as a major improvement. The two pieces of legislation—the 2006 law on concessions and the 2015 law on PPPs—have formed a sufficient base for partnership implementation in many sectors, including transport infrastructure, the social sector (hospitals, schools and kindergartens) and the energy sector. The latter naturally incorporates RES as a growing industry in which Kazakhstan has a particular interest, due to its current dependency on fossil fuels. At the same time, interviewees noted a range of conditions that are linked to the legislative and regulatory framework and are currently underdeveloped and/or lacking. There was no indication that the lacking provisions are critical to the extent that they seriously impede PPP deployment. Rather, the interviewees identified these provisions as improvement opportunities from which the overall legislative and regulatory PPP framework would benefit.

Conclusion

Comparing Kazakhstan's evolving policy on renewable energy with the policy on PPPs, it is worth emphasising that the established legislative basis for partnerships ensures their fit for the purpose, i.e. the promotion of RES. Of course, the RES sector is not the only one where partnerships could be used. PPPs may be deployed in many sectors. However, the unlimited nature of RES assures the long-term nature of the projects for producing green energy, which is a typical characteristic of an industry for PPP deployment.

The government's economic development agenda could be well-served by PPPs as they form long-term business arrangements, generate income, create jobs and, hence, contribute to the economic and social sustainability. Partnerships also have the capacity to attract private funding for the production of green energy from RES and, hence, diversify the economy. As a PPP has already been launched in the Kazakhstani energy sector (namely, the project in Northern Kazakhstan that involves power generation and transmission to the Aktobe region, which is a concession for 17 years, since 2005), this experience could form the basis for expanding PPP deployment to an emergent RES sector.

The appraisal of conditions for effective PPP implementation reveals the need to continuously improve the clarity and consistency of the legal PPP framework, simplify the procedures and requirements for the PPP formation, institute market discipline for either partner and design a transparent long-term tariff policy (Hardcastle et al. 2005; Yang et al. 2009). Although, in Kazakhstan, the legal and regulatory PPP framework is in place, certain deficiencies require the government's attention. Without trivialising them, one can consider them as improvement opportunities, or governance challenges, that do not diminish the value of PPPs as fit for purpose, i.e. for the promotion of RES.

What is the degree of PPPs' suitability for an emerging government policy that aims to promote the production and consumption of green energy in Kazakhstan? In the field of renewable energy, PPPs should be viewed not merely as electricity production plants, but also as a means to contribute to sustainability through job creation, income generation and reduction of the nation's dependency on fossil fuels. Furthermore, it is likely that PPP deployment in the RES sector will diversify the economy by expanding this emerging industry. In addition, in the long run PPP deployment would significantly reduce the governmental involvement in power generation, which means the reduction of the size and scope of the government sector and the corresponding expansion of the private sector. Therefore, in ex-Soviet nations, such as Kazakhstan, in which traditionally the government sector prevailed, PPPs may contribute to the development of private enterprise.

Nonetheless, despite being fit for purpose, PPPs as a practical tool for Kazakhstan's policy implementation require improvement. Among the areas needing government attention are risk allocation decisions that would satisfy both public and private partners; incentives for better performance and penalties for not meeting certain performance standards; and partner interaction issues. Finally, PPPs would be most effective when deployed within a broader framework: not only they would serve the policy on green energy, but also contribute to the development of clusters of entrepreneurship in which the production of power from RES forms one of the critical components. Kazakhstan has already adopted a broader framework: Strategy Kazakhstan 2050, which includes long-term goals to expand PPP deployment and increase utilisation of renewable energy sources. The alignment of RES expansion with PPPs as one of the implementation instruments might ensure a tangible contribution to the nation's sustainable development.

Acknowledgements This chapter has been previously published as: Mouraviev, N. and Koulouri, A. 2017. 'Public private partnerships as a policy tool for the sustainable utilisation of renewable energy sources: the case of Kazakhstan', in N. Mouraviev and N. Kakabadse (eds.), *Public-Private Partnerships in Transitional Nations: Policy, Governance and Praxis.* Cambridge Scholars Publishing: Newcastle upon Tyne, 220–240. Published in the current book with the permission of Cambridge Scholars Publishing.

References

Ackermann, T., Andersson, G., & Soder, L. (2001). Overview of government and market driven programs for the promotion of renewable power generation. *Renewable Energy, 22,* 197–204.

Agnolucci, P. (2006). Use of economic instruments in the German renewable electricity policy. *Energy Policy, 34,* 3538–3548.

BGR (Federal Institute for Geosciences and Natural Resources). (2013). *Energy study 2013.* Reserves, resources and availability of energy resources (17). www.bgr.bund.de/EN/Themen/Energie/Downloads/energiestudie_2013_en.pdf?__blob=publicationFile&v=2. Accessed 5 Jan 2017.

Bougrain, F. (2012). Energy performance and public private partnership. *Built Environment Project and Asset Management, 2*(1), 41–55.
Dinica, V. (2008). Initiating a sustained diffusion of wind power: The role of public-private partnerships in Spain. *Energy Policy, 36*(9), 3562–3571.
Dinica, V. (2011). Renewable electricity production costs—A framework to assist policy makers' decisions on price support. *Energy Policy, 39*, 4153–4167.
Dutra, R. M., & Szklo, A. S. (2008). Incentive policies for promoting wind power production in Brazil: Scenarios for the Alternative Energy Sources Incentive Program (PROINFA) under the New Brazilian electric power sector regulation. *Renewable Energy, 33*, 65–76.
Energy Charter Secretariat. (2013). *Investment climate and market review in the energy sector of Kazakhstan*. www.energycharter.org/what-we-do/investment/investment-climate-and-market-structure/investment-in-kazakhstan-2013/. Accessed 5 Jan 2017.
European Bank for Reconstruction and Development (EBRD). (2008). Securing sustainable energy in transition economies. www.ebrd.com/news/publications/brochures/securing-sustainable-energy-in-transition-economies.html. Accessed 5 Jan 2017.
EXPO2017 Future Energy website. https://expo2017astana.com/. Accessed 7 Jan 2017.
Government of the Republic of Kazakhstan. (2016). *Kazakhstan is creating 'Green' University*. Ministry of Energy of the Republic of Kazakhstan press release, 15/09/2016. www.publicnow.com/view/AE76956A883DBE146BD1E1F300889E03FC66D07B?2016-09-15-22:31:26+01:00-xxx1965. Accessed 14 Jan 2017.
Green Academy website. www.green-academy.kz/en/. Accessed 7 Jan 2017.
Green Bridge Partnership Programme website. https://sustainabledevelopment.un.org/partnership/?p=2237. Accessed 7 Jan 2017.
Green Economy Concept. (2013). *Concept for transition of the Republic of Kazakhstan to green economy*. Decree of the President of the Republic of Kazakhstan on 30 May 2013, No. 557. http://gbpp.org/wp-content/uploads/2014/04/Green_Concept_En.pdf. Accessed 14 Jan 2017.
Hardcastle, C., Edwards, P. J., Akintoye, A., & Li, B. (2005). Critical success factors for PPP/PFI projects in the UK construction industry: A factor analysis approach. *Construction Management and Economics, 23*(5), 459–471.

International Energy Agency (IEA). (2014). *Eastern Europe, Caucasus and Central Asia. Energy highlights. Energy policies beyond IEA countries.* www.iea.org/publications/freepublications/publication/energy-policies-beyond-iea-countries—eastern-europe-caucasus-and-central-asia-2015–highlights.html. Accessed 24 Dec 2016.

International Energy Agency (IEA). (2015). *Eastern Europe, Caucasus and Central Asia. Energy policies beyond IEA countries.* www.iea.org/publications/freepublications/publication/IDR_EasternEuropeCaucasus_2015.pdf. Accessed 5 Jan 2017.

International Energy Agency (IEA). (2016). *Kazakhstan balances for 2014.* www.iea.org/statistics/statisticssearch/report/?country=Kazakhstan&product=balances. Accessed 11 Jan 2017.

Jamali, D. (2004). A public-private partnership in the Lebanese telecommunications industry: Critical success factors and policy lessons. *Public Works Management and Policy, 9*(2), 103–119.

Kazakhstan 2050. (2012). *Strategy Kazakhstan-2050: New political course of the established state.* Address by the President of the Republic of Kazakhstan, Leader of the Nation, N. Nazarbayev: December 14, 2012. https://strategy2050.kz/en/multilanguage/. Accessed 5 Jan 2017.

Kazakhstan Public-Private Partnership Centre. (2016). http://kzppp.kz. Accessed 15 Jan 2017.

Land Code of the Republic of Kazakhstan. (2003). *Code of the Republic of Kazakhstan of 20 June 2003, No. 442.* http://adilet.zan.kz/eng/docs/K030000442_. Accessed 14 Jan 2017.

Law of the Republic of Kazakhstan on Support of the Use of Renewable Energy Sources. (2009). *Law of the Republic of Kazakhstan of 4 July 2009, No. 165-IV ZRK.* http://cis-legislation.com/document.fwx?rgn=28433. Accessed 5 Jan 2017.

Liming, H., Haque, E., & Barg, S. (2008). Public policy discourse, planning and measures toward sustainable energy strategies in Canada. *Renewable and Sustainable Energy Reviews, 12,* 91–115.

Martins, A. C., Marques, R. C., & Cruz, C. O. (2011). Public-private partnerships for wind power generation: The Portuguese case. *Energy Policy, 39*(1), 94–104.

Menz, F. C., & Vachon, S. (2006). The effectiveness of different policy regimes for promoting wind power: Experiences from the states. *Energy Policy, 34,* 1786–1796.

Mouraviev, N., & Kakabadse, N. K. (2016). Conceptualising public-private partnerships: A critical appraisal of approaches to meanings and forms. *Society and Business Review, 11*(2), 155–173.

Mouraviev, N., & Kakabadse, N. K. (2017). *Public-private partnerships: Policy and governance challenges facing Kazakhstan and Russia*. London: Palgrave Macmillan.

Mouraviev, N., Kakabadse, N. K., & Robinson, I. (2012). Concessionary nature of public-private partnerships in Russia and Kazakhstan: A critical review. *International Journal of Public Administration, 35*, 410–420.

Pongsiri, N. (2002). Regulation and public-private partnerships. *International Journal of Public Sector Management, 15*(6), 487–495.

REEEP (Renewable Energy and Energy Efficiency Partnership). (2014). *Kazakhstan (2014)*. www.reeep.org/kazakhstan-2014. Accessed 5 Jan 2017.

Reiche, D., & Berchberge, M. (2004). Policy differences in the promotion of renewable energies in the EU member states. *Energy Policy, 32*, 843–849.

Samruk-Green Energy website. www.samruk-green.kz/en/company/missi-ya-i-videnie/. Accessed 11 Jan 2017.

Stenzel, T., & Frenzel, A. (2008). Regulating technological change—The strategic reactions of utility companies towards subsidy policies in the German, Spanish and UK electricity markets. *Energy Policy, 36*, 2645–2657.

United Nations Development Programme (UNDP). (2011). *Lessons learnt from the UNDP-GEF project "Kazakhstan–Wind Power Market Development Initiative"*. Prepared by Kieran Lettice, Astana, April 2011. http://www.kz.undp.org/content/dam/kazakhstan/docs/UNDP_KAZ_wind_power_market.pdf?download. Accessed 27 Dec 2016.

Valle Costa, C., La Rovere, E., & Assmann, D. (2008). Technological innovation policies to promote renewable energies: Lessons from the European experience for the Brazilian case. *Renewable and Sustainable Energy Reviews, 2*, 65–90.

Wachsmann, U., & Tolmasquim, M. T. (2003). Wind power in Brazil—Transition using German experience. *Renewable Energy, 28*, 1029–1038.

www.akorda.kz. (2010). *2010–2014 state program of the accelerated industrial and innovative development of the republic of Kazakhstan*. Decree of the President of the Republic of Kazakhstan of 19 March 2010, No. 958. www.akorda.kz/upload/SPFIID.doc. Accessed 5 Jan 2017.

www.globalmethane.org. (2013). *Action plan for the development of alternative and renewable energy in Kazakhstan 2013–2020*. Resolution No. 43 of the Government of the Republic of Kazakhstan, 25 January 2013. www.globalmethane.org/documents/GMI_KZ_Seminar_FIT_Tuleubaeva-2015-Sept.pdf. Accessed 10 Jan 2017.

www.kazembassy.org.uk. (2015). *100 concrete steps for the implementation of five institutional reforms.* http://www.kazembassy.org.uk/en/pages/page/82. Accessed 24 Dec 2016.

Yang, J., Shen, G. Q., Ho, M., Drew, D. S., & Chan, A. P. C. (2009). Exploring critical success factors for stakeholder management in construction projects. *Journal of Civil Engineering and Management, 15*(4), 337–348.

Zakon Respubliki Kazakhstan. (2006). *O kontsessiyakh* (in Russian) [The law of the Republic of Kazakhstan "On concessions"]. www.kzppp.kz. Accessed 17 Aug 2012.

Zakon Respubliki Kazakhstan. (2015). *O gosudarstvenno-chastnom partnerstve* (in Russian) [The law of the Republic of Kazakhstan "On Public-Private Partnerships"]. www.kzppp.kz. Accessed 17 Jan 2016.

5

Barriers to Energy Security in Australia: The Electricity Sector Governance and the Need for Change

Carl Tidemann

Introduction

The recent Paris Agreement enshrines the global decision to reduce greenhouse gas (GHG) emissions to limit global warming to 2 degrees Celsius or lower (UN 2018). For this reduction to occur, rapid decarbonisation in all sectors of the economy is crucial. Electricity production is responsible for nearly 40% of global emissions (IEA 2017); therefore, transitioning towards renewable sources of electricity generation (RES-E) offers high levels of mitigation potential. As a result, the share of RES-E as a proportion of global, total final energy consumption increased to 5.4% in 2016 (REN21 2018). It is also of note that renewable generation technologies have decreased in cost to the point of being less expensive than other conventional technologies (Lazard 2017).

C. Tidemann (✉)
Fenner School of Environment and Society, Australian National University, Acton, ACT, Australia
e-mail: carl.tidemann@anu.edu.au

N. Mouraviev and A. Koulouri (eds.), *Energy Security*,
https://doi.org/10.1007/978-3-030-01033-1_5

Although increasing RES-E reduces emissions, it also creates unique challenges. Solar and wind power, the technologies that have seen the highest global growth in recent years (IEA 2018a), are intermittent and can therefore cause issues related to electricity system security (Riesz and Milligan 2014). Consequently, energy security has transformed from concerns centred on the external supplies of fossil fuels beginning with the 1970s 'Oil Shocks' (Mitchell 2016), to now being 'closely entangled with other energy policy problems such as providing equitable access to modern energy and mitigating climate change' (Goldthau 2011). This has recently been embodied in the 'energy trilemma', in which energy policies need to balance the trade-offs between security (both of the electricity system and the supply of inputs), economic competitiveness and affordability, and environmental sustainability—namely GHG emissions mitigation (Ang et al. 2015).

Following the global trend, Australia has seen steady growth in its total share of RES-E. However, fossil fuel generation has rebounded and the year on year rate of growth of RES-E is decreasing (DEE 2017). Moreover, the RES-E transition has been preceded by, and is running concurrently with, four major drivers of change. The first is the move away from state-owned monopoly utilities towards a corporatised and/or privatised sector to improve efficiency in a process known as marketisation (AEMC and KPMG 2013). The second is a long-running trend in global public policy known as agencification. Governments create semi-autonomous agencies, which are intended to operate at an arm's length from the government and complete public tasks such as regulation, service delivery and policy implementation (Verhoest et al. 2012). The authors further suggest that agencification, in comparison with government bureaucracy, is intended to limit political influence. The third is an early decision in the restructuring process to make environmental decisions relating to electricity the responsibility of government (Macgill and Healy 2013). Finally, there has been a decades-long political debate over the importance of GHG mitigation and whether renewables should be integrated at all (see Lucas 2017). The split has typically been between the 'left'-leaning Australian Labour Party being for, and the 'right'-leaning Coalition (Liberal and National parties) being against ambitious mitigation (see Warren et al. 2016). Consequently, the electricity sector is now governed by a variety of federal level, quasi-governmental agencies as well as different Federal and state governments, all implementing various policies to promote environmental outcomes.

This chapter is concerned with particular facets of the expanded notion of energy security and the energy trilemma in Australia. The first is security of the electricity system, in particular its reliability, considering recent, intense political scrutiny, triggered mainly by a statewide blackout in South Australia (SA) on 28 September 2016, which the Coalition Federal Government blamed on SA's high levels of RES-E (Lucas 2017) (see section "The 2016 South Australia Blackout and Ensuing Reform Agenda"). The second facet consists of the barriers to addressing the energy trilemma, in particular, risks posed by politicisation. As the World Energy Council (2016, p. 12) suggests, 'predictable and durable energy policies that go beyond the political cycle and have clearly defined goals are the cornerstones of a sustainable energy system'. These are important issues to comprehend if there is to be an increase in RES-E use.

This chapter has been informed by 14 interviews with government officials, key industry bodies, consultants and market participants; submissions to government enquiries; and the peer-reviewed academic literature. It does not attempt to answer every possible question that could be asked about modern energy security issues. Rather, the intent is more exploratory, seeking to draw out insights and possible lessons from the Australian experience for the benefit of policymakers here and elsewhere, who are contemplating future steps for governance in light of the energy trilemma. Rapid change in the sector necessitates the use of timeframes for analysis. This chapter explores the period from the September 2016 SA blackout up to the SA election in March 2018 when the Labour government, having recently implemented a variety of electricity system security measures, lost power to the (Conservative) Coalition. However, the chapter extends analysis at the Federal level to June 2018 and includes negotiations surrounding the 'National Energy Guarantee'—the Federal Government's new climate and electricity policy platform (ESB 2018).

The chapter is structured as follows. Section "The Australian Electricity Sector" outlines the current electricity system context of Australia's

National Electricity Market (NEM) and highlights major issues based on secondary data analysis. Section "Governance and the National Electricity Market" explores the current governance of the NEM. Section "The 2016 South Australia Blackout and Ensuing Reform Agenda" explores ways in which different levels of the Australian government are trying to address the energy trilemma, in particular actions by the Federal and SA governments following the 2016 blackout. Section "The Coal Agenda" discusses the broader implications of change in the current system. Finally, section "The Federal Government and the National Energy Guarantee" concludes and suggests lessons for future governance arrangements.

The Australian Electricity Sector

The Australian electricity sector is dominated by the NEM (Fig. 5.1), first organised in 1998, which provides some 90% of the electricity consumed nationally and includes the eastern states of: Queensland (QLD); New South Wales (NSW), which includes the Australian Capital Territory (ACT); Victoria (VIC); Tasmania (TAS); and South Australia (SA). Each region operates as a separate, regional, spot-market and is interlinked with other regions by interconnectors. Western Australia and the Northern Territory operate on separate grids (AEMO 2017a). Analysis here is limited to the NEM due to the common electricity governance structure shared by its member states.

In 2015, the Australian electricity sector had the highest CO_2 intensity among all International Energy Agency (IEA) member countries and almost twice as high as the IEA average (IEA 2018a). This is due to a heavy reliance on abundant, and thus cheap supplies of coal, particularly brown coal (lignite)—a carbon fuel with very high emissions intensities (Table 5.1)—in the state of VIC and a small amount in SA (Table 5.2).

However, the national statistics obscure the differing generation portfolios of the states, due in a large part, to the difference in available resources. This is further compounded by the different policy instruments implemented at federal and state/territory level. SA, for example, has a very high penetration of wind power due to abundant wind

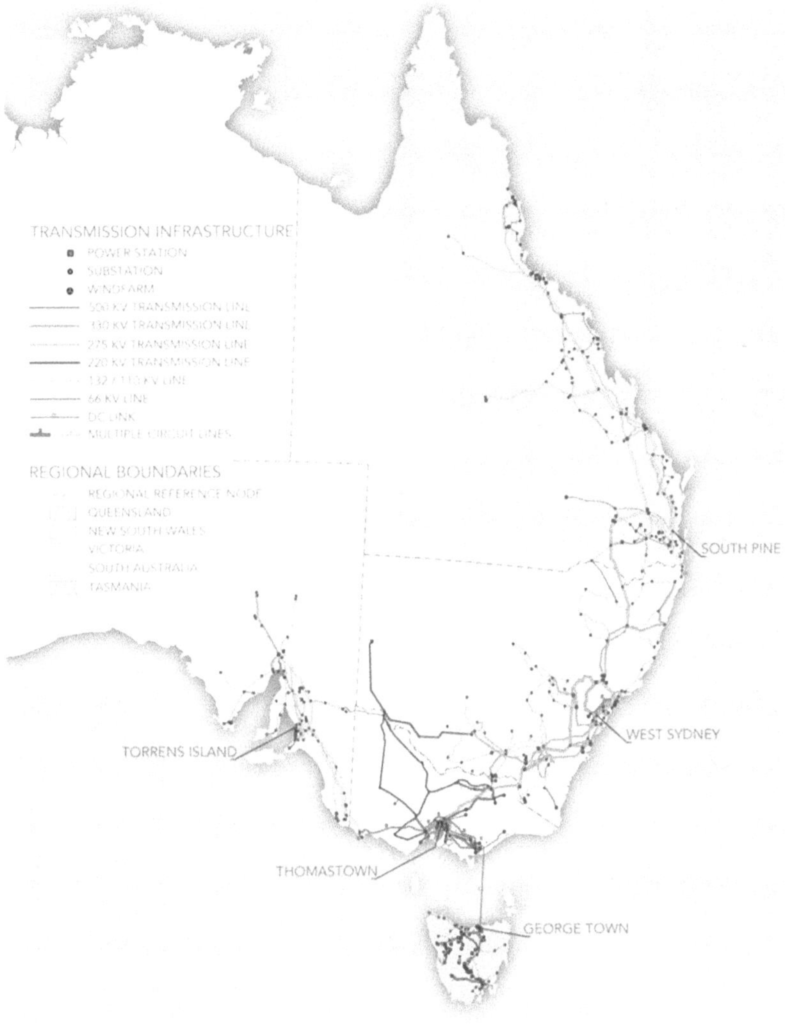

Fig. 5.1 The National Electricity Market (*Source* AEMC 2018b)

resources and the effects of the Renewable Energy Target (RET); this is explored in more detail in section "Government Policies to Promote Renewables". Table 5.2 shows the proportion of renewable generation in each region of the NEM and the total for Australia.

Table 5.1 Estimated operating emissions for new power stations

Generation and resource type	Estimated operating emissions as generated (kg CO_2e/MWh)
Subcritical brown coal	1140
Supercritical brown coal	960
Subcritical black coal	940
Supercritical black coal	860
Ultra-supercritical brown coal	845
Ultra-supercritical black coal	700
Open-cycle gas turbine	620
Combined-cycle gas turbine	370
Wind	0
Hydro	0
Solar photovoltaic	0

Source Finkel et al. (2016, p. 63)

Table 5.2 Percentage of electricity generation by resource type for regions of the National Electricity Market and Australia, 2016–2017

	Australia	QLD	NSW	VIC	SA	TAS
Non-renewable fuels						
Black coal	45.5	72.1	79.2			
Brown coal	16.8	0.0	0.0	82.5		
Natural gas	19.7	19.5	4.6	4.9	52.2	8.4
Oil products	2.4	1.6	0.4	0.3	1.0	0.2
Total non-renewable	84.4	93.1	84.2	87.7	53.2	8.6
Renewable fuels						
Biomass	1.4	2.5	1.2	1.3	0.8	0.3
Wind	4.8	0.0	2.6	6.8	37.4	10.2
Hydro	6.4	1.0	8.7	1.8	0.0	79.8
Large-scale solar PV	0.3	0.0	0.8	0.0	0.0	0.0
Small-scale solar PV	2.8	3.3	2.4	2.3	8.6	1.1
Geothermal	0.0	0.0	0.0		0.0	0.0
Total renewable	15.6	6.9	15.8	12.3	46.8	91.4

Source Department of Environment and Energy (2017), Australian Energy Update Table O

The Australian electricity sector is vital to the economy, having contributed 22.7 billion AUD or 1.4% of Gross Value Added to the Australian economy in 2017 (ABS 2018). Australia is, however, an outlier among industrialised developed countries, as its economic structure and wealth are highly dependent on energy and commodity exports

(Hermanns 2015, p. 110). Its energy and commodity producers are heavily subsidised, mostly in the form of tax concessions. The OECD (Organisation for Economic Co-operation and Development 2018) reports over 9 billion AUD in budgetary transfer and tax expenditure, such as fuel tax credits and direct funding of coal exploration. Using a more expansive methodology, the International Monetary Fund (IMF 2015) puts this figure at over 30 Billion USD. Hermanns (2015) further suggests that export competitiveness and job security in these industries are strengthened by low electricity prices. Unsurprisingly, these industries want to protect their interests and thus expend significant resources lobbying different governments. These efforts have led to major policy modification at Australian federal level in particular, including generous free permit allocations to a variety of emissions intensive and trade-exposed industries under the now-repealed carbon pricing scheme (Pezzey et al. 2010), and the failure of the Mining Resources Rent Tax due to a strong, public campaign by the mining industry (Bell and Hindmoor 2014). Although these campaigns were not directly related to the electricity sector, they foreshadow actions related to increasing the proportion of RES-E (see section "Transforming the Institutional Structure"). The following sections introduce the current governance arrangements in the NEM.

Governance and the National Electricity Market

Current governance arrangements in the NEM follow a decades-long process of restructuring, that began with the liberalisation of the Australian economy in the beginning of the 1980s (Garnaut 1994). Energy-intensive sectors, then exposed to international competition, lobbied for a more competitive electricity supply, resulting in a move from state government-owned monopolies towards corporatisation and privatisation, and eventually the creation of the NEM in December 1998 (AEMC and KPMG 2013). In the mid-2000s, the state and territory governments operating within the recently formed NEM decided to hand over the economic regulation of the sector to the Australian Energy Regulator (AER), which saw the creation of the Australian Energy

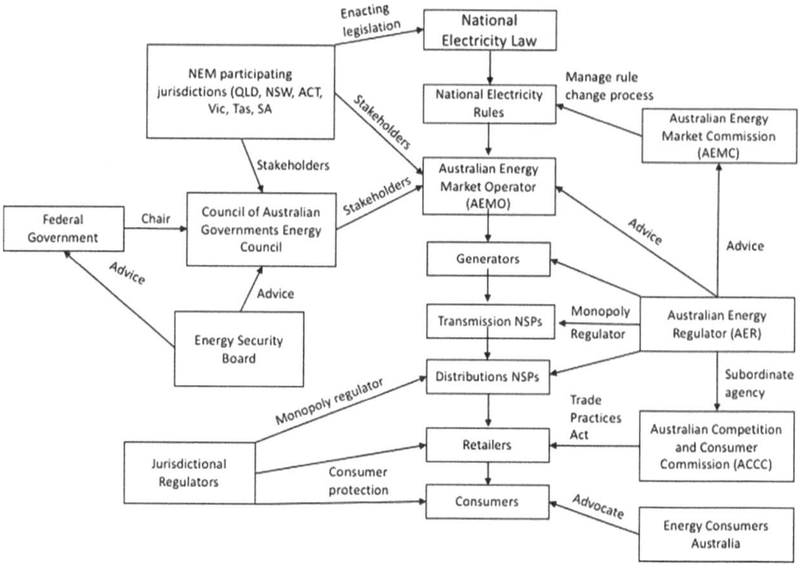

Fig. 5.2 National Electricity Market governance system post South Australian blackout (*Source* Adapted from Ball et al. (2011). *Note* NSPs = network service providers)

Market Commission (AEMC). 'This "federalisation" and bifurcation of regulation, was justified on the basis that it would lower the cost and complexity of regulation as perceived by investors, enhance regulatory certainty and ensure national uniformity' (Mountain 2014, p. 189).

Governance Structure of the National Electricity Market

The current governance system of the NEM (Fig. 5.2) was decided by negotiations between the Federal and state/territory governments. The Australian Energy Market Agreement (AEMA), which was negotiated between the Federal Government and member jurisdictions in 2004 and amended in 2013, sets out the roles and responsibilities of the actors as well as the state/territory and Federal Government's agreement to create the NEM. The AEMA sets a number of goals, although it is concerned mainly with providing affordable and reliable energy with the interests of consumers in mind. The AEMA does however mention the need to 'address greenhouse emissions from the energy sector, in light of the concerns about climate change and the need for a stable long-term framework for investment in energy supplies' (COAG Energy Council 2013).

The Federal Government's lack of jurisdiction over the electricity sector requires the National Electricity Law (NEL) to be contained in a Schedule to the National Electricity (South Australia) Act 1996 (SA). The NEL is applied as law in each participating jurisdiction by application statutes, for example, the National Electricity (VIC) Act 2005. The overarching goals of the NEM are enshrined in the National Electricity Objective (NEO), stated in the National Electricity (South Australia) Act s. 7 (NEL), which notably does not include environmental sustainability (i.e. greenhouse emissions control), namely:

> to promote efficient investment in, and efficient operation and use of, electricity services for the long-term interests of consumers of electricity with respect to:
>
> - price, quality, safety, and reliability and security of supply of electricity
> - the reliability, safety and security of the National Electricity system.

Enactment of the enabling legislation by member states transfers regulatory power to the federal-level bodies, which includes the following:

- the Council of Australian Governments Energy Council (COAG EC) has responsibility for national policy direction. This comprises energy-related ministers from each Australian state and territory and is chaired by the Federal environment or energy minister of the day. Decision-making is carried out by consensus of the members. The EC is supported by the Senior Committee of Officials, which advises the Council and develops issues for its consideration in the context of the Council's Terms of Reference and other issues as identified and agreed by ministers;
- the Australian Energy Market Commission (AEMC) is accountable to the COAG EC and is responsible for the rule-making process;

- the Australian Energy Regulator (AER), part of the Australian Competition and Consumer Commission (ACCC), enforces the rules made by the AEMC; and
- the Australian Energy Market Operator (AEMO) operates the market as well as having responsibility for the functioning of the electricity system and the 'coordination of the strategic development of the National Electricity grid' (AEMC 2018c).

A number of other organisations exist within the NEM structure, including Energy Consumers Australia, the ACCC and state/territory jurisdictional regulators. However, these are not relevant to the analysis here.

Macgill and Healy (2013, p. 616) highlight the 'early decision by Australian policymakers, seen also in many other jurisdictions, to focus electricity industry restructuring policy on improving economic efficiency through competitive pressures, while pursuing environmental objectives through "external" policy efforts'. Given this decision, the main paradigms currently driving the governance of the electricity sector are competitiveness and the security of the system. This focus on only two aspects of the energy trilemma is important when considering the integration of RES-E and will be discussed in section "The Finkel Review". Moreover, the separation of environmental objectives results in a variety of different policies being implemented by state/territory and Federal governments.

Government Policies to Promote Renewables

The desire to reduce emissions through the use of RES-E has resulted in the previous Federal Labour government (in power 2007–2013) and all state and territory governments implementing policies to promote their use. This, like the political debate outlined in section "Introduction", was, and still is, highly dependent on the political party in power at the time.

The previous (Labour) Federal Government, acting to reduce emissions under the Kyoto Protocol, implemented a variety of climate and energy policies, including a RET. The RET, made up of large and small-scale components, is a quota obligation that is fulfilled by a requirement for retailers to source tradable emission certificates (IEA 2018b). Following the election of the (Coalition) Federal Government and subsequent RET review in 2014, the target was reduced from 41,000 gigawatt-hours (GWh) in 2015 to 33,000 GWh in 2020, which led to a period of industry uncertainty and reduced investment in RES-E (IEA 2018b).

This period of uncertainty, coupled with the removal of other federal-level emissions mitigation policies (discussed in section "The Finkel Review"), has led most state governments to implement their own RES-E targets and policy instruments, which are summarised in Table 5.3.

The 2016 South Australian Blackout and Ensuing Reform Agenda

On 28 September 2016, the entire state of SA experienced a blackout, following severe weather. The intense media debate that followed saw the Federal Government blame the blackout on SA's high penetration of renewables, particularly wind energy generation (see Table 5.2 and Lucas 2017). However, the cause, explained in a series of reports from the Australian Energy Market Operator, was that a series of events including tornadoes, collapsed transmission lines, the overloading of the interconnector with VIC and a lack of fossil fuel and renewable capacity led to the blackout (AEMO 2017c). Prior to the event, the operation of the NEM, particularly the security of the system, as outlined in the National Electricity Rules (AEMC 2018c), was the responsibility of the Australian Energy Market Operator (AEMO) and the AEMC's Reliability Panel, which set system security standards and guidelines (AEMC 2018d). Chief among these standards is a reliability standard that dictates:

> for generation and inter-regional transmission elements in the National Electricity Market is a maximum expected unserved energy (USE) in a region of 0.002% of the total energy demanded in that region for a given financial year. (AEMC 2018c, p. 162)

This means that for any region, for example, NSW, electricity supply should be maintained at 98.998% of that year's demand. The exact

Table 5.3 Summary of federal and state/territory RES-E targets and policy instruments

Jurisdiction and government	Target	Policy instruments
Federal Labour	33 terawatt hours (TWh) of renewable energy (or 23.5% of the electricity mix) by 2020	Quota and tradable certificate scheme—responsibility for retailers to procure generation or certificates
Australian Capital Territory	100% by 2020 (legislated)	Feed in Tariffs for small-scale solar PV (FiTs), reverse auctions (RA), battery subsidies
Queensland—Labour	50% by 2030 (unlegislated)	FiTs, RA for 150 MW of solar, further RA for 400 MW including 100 MW of storage
Victoria—Labour	25% by 2020, 40% by 2025 (legislated)	FiTs, proposed RAs
New South Wales—Coalition	Supports national plan of 20% by 2020	FiTs, tender announced for 137 GWh electricity for rail project
South Australia—Current Coalition (targets and policies set by previous Labour Government)	50% by 2025 (53% achieved in 2017)	FiTs
Tasmania—Liberal	No target—high penetration related to hydropower development	FiTs

Source Adapted by the author from IEA (2018a)

causes of the blackout are in many ways irrelevant, considering its use for justifying the governance changes that followed. It is, however, of note that the reliability standard for SA was not affected, because the blackout was considered an event that could not have been foreseen (AEMC 2017). Following the blackout, a number of actions took place relating to the energy trilemma, which will be discussed below. These include the changes to the current governance system and the reform agenda of the Federal and SA governments in particular.

Reviews of the Electricity Sector

Prior to the SA blackout, many reviews of the electricity sector were conducted, which included aspects of the way the sector is and was governed, the security of the system, affordability and the integration of renewables. A complete summary is beyond the scope of this chapter, though the Productivity Commission's Inquiry into the Regulation of Electricity Networks (2013) and COAG Energy Council's Review of Governance Arrangements (the Vertigan Review) (2015) in particular, are relevant to the analysis.

The Finkel Review

At a 6 October 2016 meeting, the COAG Energy Council agreed to an 'Independent Review into the Future Security of the National Electricity Market'—commonly known as the Finkel Review (Finkel et al. 2016). The review was to investigate a broad range of issues including market design, increasing electricity prices, emissions mitigation, governance of the sector, integration of variable renewables, technology transformation and the role of consumers.

Prior to the blackout, AEMO already had a system security investigation process underway (the Future Power System Security Program) (AEMO 2017b), and the AEMC was undertaking a System Security Market Frameworks Review that would address the issues identified by AEMO (AEMC 2018d). Nevertheless, these work programmes, and the fact that the reliability standard remained intact, did not prevent the (Coalition) Federal Government moving forward. What followed was a series of events that have shaped the Federal Government's reform agenda, which will be explored in the following section. Section "The Finkel Review" explores some of the main focus points of submissions to the Finkel Review, which suggested changes to the governance structure outlined above.

The Finkel Review, released in June 2017, made fifty recommendations across a range of issues related to the security of the system, emissions mitigation and the governance of the electricity sector (Finkel et al. 2017). Following its release, the Federal Government agreed to 49 of the 50 recommendations. The notable exception was the policy instrument—an emissions intensity scheme or clean energy target—that Finkel suggested to reduce emissions in the electricity sector. Instead, the government, having sought advice from the newly formed Energy Security Board (Finkel recommendation 7.2) proposed the National Energy Guarantee (NEG). The exact details of the policy still remain unclear, though it combines an emissions baseline and a generator reliability standard (Finkel recommendation 3.3).

The Finkel Review also outlined issues highlighted in the Productivity Commission and Vertigan Reviews. The Productivity Commission (2013, p. 36) suggested that:

> reform appears to have been frustrated by complex processes, constant and overlapping reviews, and a lack of agreement by relevant governments about either the reforms themselves or the need for more timely progress to a genuinely NEM-wide approach to energy regulation.

Furthermore, it suggested that NEM governance processes often descend into 'paralysis by analysis' (Productivity Commission 2013, p. 36). With respect to the Vertigan Review, Finkel et al. (2017) suggested that although the review found the governance of the NEM to be fundamentally sound, there remain a number of the recommendations that are yet to be implemented:

- the appointment of additional AEMC and AER commissioners
- expedited rule change processes

- a comprehensive review of the Rules
- structural separation of AER from the ACCC
- a new Statement of Role for AEMO.

Transforming the Institutional Structure

The Finkel Review submissions and subsequent recommendations highlight a variety of issues. However, there are two long-running (they were also identified and addressed by the Vertigan Review in 2015 [Vertigan et al. 2015]) challenges identified in submissions to the Finkel Review.

The first challenge is the lack of environmental consideration in the National Electricity Objective (NEO), noted above in section "Governance Structure of the National Electricity Market". Integrating environmental concerns with competitiveness and security was widely identified in submissions (to both the Vertigan and the Finkel reviews) as the best opportunity to integrate emissions considerations into the functioning of the sector. Enhancing the NEO to include environmental concerns would address the energy trilemma, though notably, this was not one of the Finkel recommendations. The NEO is important because it dictates the way in which the AEMC approaches rule changes. Without any environmental imperative, the decisions are based on competitiveness, and the security of supply of electricity and the system itself. As a result, it is unlikely that rule changes will be made to promote renewables on the grounds of emissions mitigation.

The second challenge is the COAG Energy Council's structure and workings. Under the AEMA, the COAG Energy Council has been given responsibility for facilitating a nationally consistent approach to energy policy, overarching responsibility for governance and institutional arrangements and regulatory and legislative frameworks (COAG Energy Council 2013). This is operationalised in its capacity to shape the way the AEMC makes rule changes (as well as being able to make rule change requests) and to recommend appointments of commissioners to the AEMC and members to the AER (COAG Energy Council 2013). However, with respect to changes suggested by the 2015 Vertigan Review, which are yet to be implemented (see section "Reviews of the Electricity Sector"),

Energy Networks Australia, an industry body that represents electricity network businesses, suggested that:

> regardless of the merits of any single measure, this delay [in implementing suggestions from the Vertigan Review] suggests that the existing strategic policy, governance and implementation process within the COAG Energy Council itself is not delivering outcomes efficiently and effectively. (ENA 2017, p. 52)

Furthermore, the Vertigan Review itself recommended that:

> the Council and Senior Council of Officials appear to lack a focus on strategic direction and are therefore not providing effective and active policy leadership to the energy sector. Whilst the inherent structure of the Council cannot be altered, the Council can improve the visibility, transparency and accountability of its processes and operations to more effectively progress strategic energy market reform. (Vertigan et al. 2015, p. 7)

Considering the Finkel Review did not mention changing the National Electricity Objective, changes to the functioning of the COAG EC assume that the emissions mitigation targets or policies set by the Federal Government will be of sound design and ambition. However, as Rogelj et al. (2016) outline, the pledges made under the Paris Agreement, Australia's included, are not enough to limit global warming to 2 degrees or less. Changes to the governance structure, though important, would then seem redundant in promoting more RES-E, given the Federal Government's reform agenda, which will be discussed in the following sections.

The Coal Agenda

As Lucas (2017) outlines, the SA blackout and subsequent release of the Finkel Review led to a highly partisan political debate that centred mainly on the use of coal as an electricity source, as opposed to the move towards renewable sources. The debate perhaps peaked on 9 February 2017, in a now infamous scene when the Federal Coalition

Treasurer, Scott Morrison, brought a piece of coal into Australian Parliament, exclaiming 'this is coal. Do not be afraid. Do not be scared. It will not hurt you' (Parliamentary Hansard 2017). Lobbying has clearly played a role in shaping the Australian climate and energy policies previously (Pezzey et al. 2010) and some aspects relating to the electricity sector will be outlined below.

Cost Discourse

Since its election in 2013, the (Coalition) Federal Government has tried, and in most cases succeeded, to dismantle the emissions and renewables policies of the previous, Labour government (see section "Government Policies to Promote Renewables"). These actions were spurred on by lobbying efforts, and used the discourse that renewable sources of electricity generation were, among other factors, not cost-competitive (Curran 2012). However, the cost of many renewable generation technologies has recently fallen below, or within the range of, all fossil fuel technologies (Lazard 2017). This fact is recognised by industry and was highlighted in the preliminary report of the Finkel Review (2016, p. 8):

> Owner-investors are exiting emissions intensive power stations as these reach the end of their design lives. It has been clear from our consultations that no one is contemplating investing in new ones, nor would financial institutions provide finance.

Consequently, this 'cost discourse' shifted to arguments around reliability and security.

Technology Neutrality

One of the preliminary Finkel Review questions related to the need for 'high efficiency low emissions' (HELE) coal-fired generation in the future (Finkel et al. 2016). Although it is not widespread in submissions, there is a concentrated effort from key organisations to promote both technology neutrality *and* the consideration of coal-fired generation. Technology neutrality is an oft-cited consideration in the promotion of technology by governments; 'picking winners' should be avoided to allow the market to decide on the most efficient technologies (Hoppmann et al. 2013). Yet, as Tidemann et al. (2018) have shown, technology neutrality related to system security, and renewable energy technology integration more broadly, is a questionable paradigm to aid electricity sector transformation. Given the following quotes from coal lobby submissions, it is also apparent that technology neutrality is being invoked, not to promote the most efficient option, but rather to allow for the promotion of technologies that may no longer be able to compete on their merits.

> Looking ahead we need to ensure there are no policy impediments for the uptake of high efficiency low emissions (HELE) coal-fired plants and carbon capture and storage (CCS). (Australian Coal Association Low Emissions Technologies 2017, p. 2)

With respect to coal:

> it is critical that all options are left on the table and that the focus remains on achieving reliable, secure, affordable electricity that over time reduces emissions. (NSW Minerals Council 2017, p. 1)

> Technology neutrality must be a central tenet of energy policy…. (Minerals Council of Australia 2017a, p. 2)

> The final report must make a clear and unambiguous statement in favour of a technology neutral approach to policy support for both CCS and HELE coal technologies. (Minerals Council of Australia 2017a, p. 3)

These submissions were coupled with the Minerals Council of Australia's TV campaign that 'highlights the role that high efficiency, low emission (HELE) coal-fired generation plants provide in reducing emissions' (MCA 2017b). However, like the cost discourse invoked before it, given the emissions profiles of different technologies shown in Table 5.1, even these new technologies cannot be considered to be low in emissions.

The actions of these lobby groups are perhaps evident in the policy principles outlined for the National Energy Guarantee relating to the emissions reduction portion of the policy.

> The emissions requirement is technologically-neutral. This means it supports the lowest-cost ways of meeting the emissions requirement, whether that is improving the efficiency of coal generators, fuel switching from coal to gas and/or building renewable energy capacity, even under different assumptions about the future (for example, assumptions about future gas prices or technology costs). (ESB 2017, p. 31)

Strength of the Coal Lobby, as Shown by the Liddell Power Station Case

Lobbying efforts, or at least the ideological stronghold of coal in Australia, are evident in recent developments in the sector related to an ageing coal-fired generator—the Liddell Power Station, NSW. Owned by one of the largest energy companies in Australia, AGL, the generator is reaching the end of its useful life and in 2015 AGL announced the plant would close in 2022. The company recently performed modelling to replace the power station with a variety of different technologies including:

- two new gas-fired generators
- 1.6 GW of renewable generation
- demand response measures
- a large-scale battery
- the upgrade of another AGL-owned coal generator
- the conversion of the Liddell generators into synchronous condensers, which aid in system security (AGL 2017).

'An assessment of AGL's plan found the replacement generation is more affordable at \$83/MWh, compared with extending Liddell at \$106/MWh. The plan was also found to deliver reliable, dispatchable power for longer, due to a longer asset life of 15–30 years, compared with a Liddell extension of five years' (AGL 2017). Nonetheless, there have still been calls from the current Federal Government for AGL to sell the coal-fired power station for the buyer to continue its operation for the five years beyond 2022 (Dziedzic 2018).

South Australia

Due to the 2007–2013 (Labour) Federal Government's RET, coupled with the state's plentiful wind resources (Coppin et al. 2003) and favourable project planning policies (Nelson and Orton 2016), SA has the highest penetration of intermittent renewables, primarily wind power, of any region of the NEM (Fig. 5.2).[1] Following the blackout, the (Labour) SA government released a policy statement that suggested a variety of measures to further integrate, and address issues relating to, RES-E. Citing the recent closure of a number of fossil-fuelled power stations, its plan (Government of South Australia 2017, p. 7) suggested that; 'without clear national policy settings there has been little to no investment to replace the thermal generation that has exited the system'.

Although there are similarities with the recommendations of the Finkel Review, the SA government's plan was much more embracing of modern technologies. It outlined a variety of measures to ensure the reliability of the system including:

- funding for battery storage, which saw the installation in late 2017 of the world's largest battery paired with a windfarm (Hornsdale Power Reserve 2018);
- a state-owned gas-fired power station;
- incentivising sourcing of gas from SA to replace coal-fired generation imported from VIC;
- an energy security target, not dissimilar to that recommended by the Finkel Review and found in the Federal Government's National Energy Guarantee, though with a stronger focus on renewable

[1]Though the Australian Capital Territory has a 100% by 2020 renewable energy target, much of the generation is outside of territory borders (ACT Government 2016).

technologies. This would have required any new generator to also provide system security benefits; and
- enacting legislation that would give SA the power to overturn decisions of AEMO if electricity shortfalls were to occur within the SA region of the NEM (Government of South Australia 2017).

It is beyond the scope of this chapter to perform a policy analysis of this plan and, following the loss of power by the Labour party in March 2018, there are few details other than the initial policy statement to consider. However, these measures are much more in line with the energy trilemma than Federal policy, given their move away from heavily polluting and higher cost fossil fuels. SA was, however, admonished by the Federal Government for 'going it alone' (Gartrell 2017). Nonetheless, the SA Government took action when they needed to in a way that suited the context of their generation portfolio. Rather than waiting for the Federal Government and COAG process to resolve, which at the time of writing has not occurred, the state used its legislative power to plan for, and in some cases implement, ambitious and technologically advanced solutions to its system security problems to allow further integration of RES-E. Unfortunately, the measures that are yet to be implemented are at risk due to a change in government following the March 2018 election.

The Federal Government and the National Energy Guarantee

Following the partial dismantling of the 2007–2013 (Labour) Federal Government's suite of climate and energy policies, the (Coalition) Federal Government is attempting to implement its National Energy Guarantee. As outlined in section "Transforming the Institutional Structure", much of the argument has centred on the cost of renewables and the need for 'technology neutrality'. Given the rapidly decreasing price of renewables and the spurious arguments for inclusion of coal under the guise of technology neutrality, the discourse has required change. The 2016 SA blackout enabled the Federal Government to 'securitise' the debate and shift focus onto the reliability issues of renewable electricity generation. Securitisation 'licenses renewed state oversight of, and involvement in, decisions about the production, transportation and consumption of energy in countries... where the state has historically unwound itself from ownership and administration of the energy system' (Bridge 2015, p. 330). This has allowed the Federal Government to create a perceived mandate to implement policy that involves emissions and security/reliability provisions. It is a perceived mandate because, as outlined in section "Governance Structure of the National Electricity Market", the Federal Government has power only to implement policies related to emissions mitigation.

The full details of the NEG are currently unknown, and due to the inclusion of reliability provisions, the Federal Government requires consensus at the COAG Energy Council to proceed with the plan. It is telling, however, that the Energy Security Board's own modelling of the NEG found that there would be little to no new investment in RES-E under the plan, and that any that occurred would be as a result of the remaining contracts agreed under the previous Labour government's RET (ESB 2017).

Discussion

The transition towards renewable generation in the Australian electricity sector, and in particular, following the SAblackout, have highlighted major issues with the current and future governance of the NEM, which will be outlined below.

Perhaps the clearest, though most difficult to remedy in the case of Australia, is the lack of integration of environmental concerns into the National Electricity Objective. By initially externalising environmental concerns from other drivers, the complexity that now exists in the sector is difficult to overcome. This has been further complicated by agencification and Australia's federal system of government. The number of organisations, agreements and legislative and regulatory instruments needed to create the governance system and allow its functioning creates overlap in responsibilities and a lack of consistent goals for the sector, as highlighted by the Productivity Commission's (2013) 'paralysis by analysis' remark. The changes to the governance system, suggested in the Productivity Commission, Vertigan and Finkel reviews, need to be implemented. In particular, the roles and responsibilities of different organisations should be confirmed to ensure consistency.

Governance problems are also compounded by the COAG Energy Council's desire for a nationally consistent approach. Although national consistency is promoted by the Finkel Review, the politically driven motivations of the current Federal Government, and their reform agenda currently being explored by the Energy Council, would suggest that a nationally consistent plan is only useful if it addresses the energy trilemma. Due to the possible inclusion of coal generation in the NEG, this is questionable because adding new coal to the energy mix will not meet emissions reduction criteria. Moreover, the various state and territory governments' ambitious energy plans, in opposition to the current Federal Government's lock-into fossil-fuelled electricity generation due to lobbying, would challenge the possibility for a nationally consistent approach. As suggested in one submission to the Finkel Review, 'the Australian energy markets require guidance that is removed from the political cycle to ensure long term certainty and to ensure political bias is minimised' (Delta Electricity 2017). It is, nonetheless, unlikely that the fossil fuel lobby in Australia and abroad will radically change their position. For rapid integration of RES-E to have any chance in the future, there is a need to address this issue. Recent research has suggested that a politically acceptable energy transition could occur by using specific market mechanisms to promote the market exit of brown coal generators (Jotzo and Mazouz 2015), but perhaps more importantly are methods for transitioning energy systems, and economies, away from fossil fuels (see for example the recent announcement of the 'Energy Transitions Hub', Energy Transition Hub 2018).

Our analysis also highlights a conflict between paradigms of change in the sector: between the Federal Government's desire for a nationally consistent approach, and the continuing agencification of the sector. Agencification is still occurring, as evidenced by the creation of the Energy Security Board. However, in the face of global crises (security, environmental and economic) and the increasing demand by citizens for integrated service delivery, a countertrend can be observed in many countries in which politicians aim to restore central control and coordination (Verhoest et al. 2012). Both agencification and the jurisdictional overreach by the Federal Government, however, represent forms of centralised control that no longer work—one by a single actor, and the other by a federalised system that arguably disallows the actions of state governments to address the issues that are unique to the context of individual regions of the NEM. As Energy Networks Australia (2017, p. 51) propose,

> the diversity of conditions, market developments, emerging demand patterns across Australian States and Territories would make assignment of the overall role to a single body to optimise outcomes across the entire electricity system infeasible.

Accordingly, the actions of the previous SA government and other current state and territory governments represent coherent and adaptive reforms that match the changing nature of the electricity sector. These cases present many opportunities for policy learning to occur at the federal level. These lessons could be integrated into a more collaborative COAG Energy Council process, rather than the Federal Government prescribing policy platforms and then seeking approval, as is the case with the National Energy Guarantee.

Conclusion

The chapter has explored the transformation of the Australian NEM governance structure through the lens of the energy trilemma: economic affordability, energy security and environmental sustainability. The analysis highlights that once-centralised electricity sectors, in terms of generation *and* governance, are not leading to the rapid integration of renewable sources of generation that are needed to reduce emissions from the electricity sector. Perhaps the starkest lesson for jurisdictions yet to restructure, or currently restructuring their electricity sector, is that the early integration of environmental, economic and energy security concerns within governance structures is of the utmost importance. The complexity that arises when governments aim to adapt already existing systems to address environmental concerns highlights this.

The need to drastically reduce emissions is changing the forms of governance that are needed. Previous arrangements, typified by agencification, centralisation and the original overarching goal of economic efficiency, are no longer sufficient. Though some aspects of top-down governance are necessary, such as national emissions targets, there is also the necessity to allow for different scales of governance that address the problems at the appropriate level and context—for example, the differing capacities of renewable generation in the NEM. It is also imperative that the concerns of the energy trilemma, and the need for sustainable development, are balanced against each other rather than allowing old paradigms of economic efficiency to dominate. This is particularly important in the setting of overarching targets for emissions reduction or renewable energy integration, which need to be ambitious and reflect current pathways to decarbonisation.

Finally, consideration must be paid to the power of lobby groups. It is beyond the scope of this chapter to fully analyse the best routes to overcome this problem. This is certainly an area for future research. Nonetheless, policymakers need to be aware of the way in which lobbying efforts use information to justify their cause. If technology neutrality is to be included in policy implementation, then it should be addressed correctly and allow technologies to compete on their merits, rather than to promote particular technologies based on vested interests.

References

ACT Government. (2016). *Renewable energy target legislation and reporting*. Website. http://www.environment.act.gov.au/energy/cleaner-energy/renewableenergytarget,-legislation-and-reporting. Accessed 28 June 2018.

AGL. (2017). *AGL announces plans for Liddell Power Station*. AGL. https://www.agl.com.au/about-agl/media-centre/asx-and-media-releases/2017/december/agl-announces-plans-for-liddell-power-station. Accessed 23 May 2018.

Ang, B. W., Choong, W. L., & Ng, T. S. (2015). Energy security: Definitions, dimensions and indexes. *Renewable and Sustainable Energy Reviews, 42,* 1077–1093.

Australian Bureau of Statistics. (2018). *5204.0—Australian system of national accounts, 2016–2017.* http://www.abs.gov.au/AUSSTATS/abs@.nsf/DetailsPage/5204.02016-17?OpenDocument. Accessed 27 June 2018.

Australian Coal Association Low Emissions Technology. (2017). *Submission to the independent review into the future security of the National Electricity Market.* Canberra: Department of the Environment and Energy.

Australian Energy Market Commission. (2017). *Annual market performance review 2017.* AEMC. https://www.aemc.gov.au/markets-reviews-advice/annual-market-performance-review-2017. Accessed 20 Apr 2018.

Australian Energy Market Commission. (2018a). *Developing electricity guidelines and standards.* AEMC. https://www.aemc.gov.au/our-work/developing-electricity-guidelines-and-standards. Accessed 23 May 2018.

Australian Energy Market Commission. (2018b). *National Electricity Market.* https://www.aemc.gov.au/energy-system/electricity/electricity-system/national-electricity-market. Accessed 22 May 2018.

Australian Energy Market Commission. (2018c). *National electricity rules—Current.* https://www.aemc.gov.au/regulation/energy-rules/national-electricity-rules/current. Accessed 1 July 2018.

Australian Energy Market Commission. (2018d). *National energy governance.* https://www.aemc.gov.au/regulation/nationalenergygovernance. Accessed 28 June 2018

Australian Energy Market Commission. (2018e). *System security market frameworks review.* https://www.aemc.gov.au/markets-reviews-advice/system-security-market-frameworks-review. Accessed 1 July 2018.

Australian Energy Market Commission & KPMG. (2013). *National Electricity Market—A case study in successful microeconomic reform.* Melbourne: KPMG.

Australian Energy Market Operator. (2017a). *AEMO publishes final report into the South Australian statewide power outage.* Australian Energy Market Operator. https://www.aemo.com.au/Media-Centre/AEMO-publishes-final-report-into-the-South-Australian-state-wide-power-outage. Accessed 1 July 2018.

Australian Energy Market Operator. (2017b). *Future power system security program—Reports and analysis.* Australian Energy Market Operator. https://www.aemo.com.au/Electricity/National-Electricity-Market-NEM/Security-and-reliability/FPSSP-Reports-and-Analysis. Accessed 1 July 2018.

Australian Energy Market Operator. (2017c). *National Electricity Market*. Australian Energy Market Operator. https://www.aemo.com.au/Electricity/National-Electricity-Market-NEM. Accessed 1 Nov 2017.

Ball, B., Ehmann, B., Foster, J., Froome, C., Hoegh-Guldberg, O., Meredith, P., … & Wagner, L. (2011). *Delivering a competitive Australian power system—Part 1: Australia's global position*. Brisbane: The University of Queensland, Global Change Institute.

Bell, S., & Hindmoor, A. (2014). The structural power of business and the power of ideas: The strange case of the Australian mining tax. *New Political Economy, 19*(3), 470–486.

Bridge, G. (2015). Energy (in)security: World-making in an age of scarcity. *Geographical Journal, 181*(4), 328–339.

COAG Energy Council. (2013). *Australian energy market agreement (as amended December 2013)*. Energy Council. http://www.coagenergycouncil.gov.au/publications/australian-energy-market-agreement-amended-december-2013. Accessed 20 Apr 2018.

Coppin, P. A., Ayotte, K. A., & Steggel, N. (2003). *Wind resource assessment in Australia—A planner's guide*. http://www.soe-townsville.org/windmonitoring/coppin_2003a.pdf.

Curran, G. (2012). Contested energy futures: Shaping renewable energy narratives in Australia. *Global Environmental Change, 22*(1), 236–244. http://dx.doi.org/10.1016/j.gloenvcha.2011.11.009.

Delta Electricity. (2017). *Submission to the independent review into the future security of the National Electricity Market*. Canberra: Department of the Environment and Energy.

Department of Environment and Energy. (2017). *Australian energy update 2017*. https://www.energy.gov.au/publications/australian-energy-update-2017. Accessed 17 Apr 2018.

Dziedzic, S. (2018). *Frydenberg pressures AGL to sell Liddell Power Station to Alinta*. http://www.abc.net.au/news/2018-04-05/liddell-power-station-government-pressures-agl-to-sell-to-alinta/9620650. Accessed 23 May 2018.

Energy Networks Australia (ENA). (2017). *Submission to the independent review into the future security of the National Electricity Market*. Canberra: Department of the Environment and Energy.

Energy Security Board. (2017). *Energy Security Board advice*. Canberra: Energy Security Board.

Energy Security Board. (2018). *Energy Security Board advice: The national energy guarantee*. Canberra: Energy Security Board.

Energy Transition Hub. (2018). *University of Melbourne-ANU to lead new energy transition hub*. http://energy-transition-hub.org/. Accessed 29 June 2018.

Finkel, A. D., et al. (2016). *Independent review into the future security of the National Electricity Market—Preliminary report*. Canberra: Department of the Environment and Energy.

Finkel, A. D., et al. (2017). *Independent review into the future security of the National Electricity Market*. Canberra: Department of the Environment and Energy.

Garnaut, R. (1994). Trade liberalization and the Washington consensus in Australia. In J. Williamson (Ed.), *The political economy of policy reform* (pp. 51–72). Washington: Institute of International Economics.

Gartrell, A. (2017). *Josh Frydenberg warns states against going it alone on clean energy target*. https://www.smh.com.au/politics/federal/josh-frydenberg-warns-states-against-going-it-alone-on-clean-energy-target-20170713-gxafok.html. Accessed 25 May 2018.

Goldthau, A. (2011). Governing global energy: Existing approaches and discourses. *Current Opinion in Environmental Sustainability, 3*(4), 213–217. http://dx.doi.org/10.1016/j.cosust.2011.06.003.

Government of South Australia. (2017). *Our energy plan SA*. http://ourenergyplan.sa.gov.au/.

Hermanns, H. (2015). Ecological modernization and the development of emissions trading schemes in Australia and South Korea. *Pacific Focus, 30*(1), 103–125.

Hoppmann, J., et al. (2013). The two faces of market support—How deployment policies affect technological exploration and exploitation in the solar photovoltaic industry. *Research Policy, 42*(4), 989–1003. http://dx.doi.org/10.1016/j.respol.2013.01.002.

Hornsdale Power Reserve. (2018). Hornsdale Power Reserve. https://hornsdalepowerreserve.com.au/. Accessed 1 July 2018.

International Energy Agency. (2017). *World energy outlook 2017*. https://www-oecd-ilibrary-org.ezproxy.unal.edu.co/docserver/weo-2017-en.pdf?expires=1527952668&id=id&accname=ocid43019721&checksum=C99798D1A48500F9691F48DF602A3DBC.

International Energy Agency. (2018a). *Energy policies of IEA countries: Australia 2018 review*. France: IEA Publications.

International Energy Agency. (2018b). *Renewables 2017*. https://www.iea.org/publications/renewables2017/. Accessed 22 June 2018.

International Monetary Fund. (2015). *IMF survey: Counting the cost of energy subsidies*. http://www.imf.org/en/News/Articles/2015/09/28/04/53/sonew070215a. Accessed 13 Apr 2018.

Jotzo, F., & Mazouz, S. (2015). Brown coal exit: A market mechanism for regulated closure of highly emissions intensive power stations. *Economic Analysis and Policy, 48,* 71–81.

Lazard. (2017, November). *Levelised cost of energy analysis* (pp. 1–21). https://www.lazard.com/perspective/levelised-cost-of-energy-2017/.

Lucas, A. (2017). Confected conflict in the wake of the South Australian blackout: Diversionary strategies and policy failure in Australia's energy sector. *Energy Research and Social Science, 29,* 149–159. http://dx.doi.org/10.1016/j.erss.2017.05.015.

Macgill, I., & Healy, S. (2013). *Is electricity industry reform the right answer to the wrong question? Lessons from Australian restructuring and climate policy.* Elsevier Inc. http://dx.doi.org/10.1016/B978-0-12-397891-2.00020-1.

Minerals Council of Australia. (2017a). *Submission to the independent review into the future security of the National Electricity Market.* Canberra: Department of the Environment and Energy.

Minerals Council of Australia. (2017b). *Launch of new coal advertising campaign.* http://www.minerals.org.au/news/launch_of_new_coal_advertising_campaign. Accessed 23 May 2018.

Mitchell, C. (2016). Momentum is increasing towards a flexible electricity system based on renewables. *Nature Energy, 1*(2), 15030. http://www.nature.com/articles/nenergy201530.

Mountain, B. (2014). Independent regulation of government-owned monopolies: An oxymoron? The case of electricity distribution in Australia. *Utilities Policy, 31,* 188–196. http://dx.doi.org/10.1016/j.jup.2014.09.011.

Nelson, T., & Orton, F. (2016). Climate and electricity policy integration: Is the South Australian electricity market the canary in the coalmine? *The Electricity Journal, 29*(4), 1–7. http://dx.doi.org/10.1016/j.tej.2016.04.007.

NSW Minerals Council. (2017). *Submission to the independent review Into the future security of the National Electricity Market.* Canberra: Department of the Environment and Energy.

Organisation for Economic Cooperation and Development. (2018). *OECD. Stat—Fossil fuel support—AUS.* http://stats.oecd.org/Index.aspx?DataSetCode=FFS_AUS. Accessed 13 Apr 2018.

Parliamentary Hansard. (2017). *Hansard—House of Representatives 9/02/2017 Parliament of Australia.* https://www.aph.gov.au/Parliamentary_Business/Hansard/Hansard_Display?bid=chamber/hansardr/e5ff85d2-b96e-4e0e-8227-2131e40beaf7/&sid=0150. Accessed 9 May 2018.

Pezzey, J. C. V., Mazouz, S., & Jotzo, F. (2010). The logic of collective action and Australia's climate policy. *Australian Journal of Agricultural and Resource Economics, 54*(2), 185–202.

Productivity Commission. (2013). *Electricity network regulatory frameworks* (Vol. 1). Canberra: Productivity Commission.

REN21. (2018). *Renewables 2018 global status report*. http://www.ren21.net/wp-content/uploads/2018/06/17-8652_GSR2018_FullReport_web_-1.pdf.

Riesz, J., & Milligan, M. (2014). Designing electricity markets for a high penetration of variable renewables. *Wiley Interdisciplinary Reviews: Energy and Environment, 4*(3), 279–289.

Rogelj, J., et al. (2016). Paris agreement climate proposals need a boost to keep warming well below 2°C. *Nature, 534*(7609), 631–639.

Tidemann, C., Engerer, N., Franklin, E., Hussey, K., & Pezzey, J. C. V. (2018). Promoting behind-the-meter battery storage: Options for more effective government support and regulation. *International Journal of Technology Intelligence and Planning, 12*(1), 77–98.

United Nations. (2018). *The Paris Agreement*. UNFCCC. https://unfccc.int/process-and-meetings/the-paris-agreement/the-paris-agreement. Accessed 1 July 2018.

Verhoest, K., Van Thiel, S., Bouckaert, G., & Laegreid, P. (Eds.). (2012). *Government agencies: Practices and lessons from 30 countries*. Basingstoke: Palgrave Macmillan.

Vertigan, M., George, A. C., & Euan, Y. (2015, October). *Review of governance arrangements for Australian Energy Markets final report October 2015*.

Warren, B., Christoff, P., & Green, D. (2016). Australia' s sustainable energy transition: The disjointed politics of decarbonisation. *Environmental Innovation and Societal Transitions, 21*, 1–12. http://dx.doi.org/10.1016/j.eist.2016.01.001.

World Energy Council. (2016). World energy trilemma 2016 defining measures to accelerate the energy transition. *World Energy Council Report*, 1–113. http://www.worldenergy.org/publications/2016/world-energy-trilemma-2016-defining-measures-to-accelerate-the-energy-transition/.

6

Ukraine's Energy Security in the New Geopolitical Context

Inna Kostrytska

Introduction

Energy security and the efficiency of the energy sector are among the cornerstones of the national security of Ukraine. Independent Ukraine inherited from the Soviet Union a potent energy sector. However, Ukraine also inherited a structurally disorganised, energy- and resource-intensive economy, represented mostly by heavy industry, as well as a number of problems in the energy sector, including those related to the safety of nuclear power generation, an inefficient and unprofitable coal mining sector, a lack of diversification of power supplies, and inefficient energy consumption.

The conflict with Russia, which began in 2014, has had a significant impact on the energy sector. The conflict resulted in the annexation of the Ukrainian Crimea by Russia and loss of control over considerable parts of Donetsk and Luhansk *oblasts* (provinces), two territories, also

I. Kostrytska (✉)
Jeantet Ukraine Law Firm, Kyiv, Ukraine
e-mail: ikostrytska@jeantet.org

N. Mouraviev and A. Koulouri (eds.), *Energy Security*,
https://doi.org/10.1007/978-3-030-01033-1_6

known as Donbas, in the east of the country. Along with the Crimea, Ukraine lost its offshore fields of oil and natural gas on the Black Sea shelf, extracting facilities and other valuable energy assets and infrastructure. It also forfeited a significant part of its energy resources in Donbas, where over 80% of Ukrainian coal mining takes place, which resulted in decreased coal extraction in Donbas (NISS 2015). The rise in tensions with Russia, and Russia's actions towards arranging gas transit to Europe via alternative routes, rather than via Ukraine, may also result in Ukraine's loss of its status as a major gas transit country for Europe.

All these problems dictate certain directions of modern energy policy and a national model of energy security in Ukraine. Among the most pressing tasks are the reduction of energy dependence upon Russia, by increasing energy efficiency and diversifying energy sources; technical and legal integration into the European energy system and the energy market; and the creation of strategic reserves and the protection of strategic energy infrastructure. While these strategic priorities have called for significant attention from policymakers in recent years, the sustainable development of the energy sector, on the basis of the principles of environmental safety has not yet received proper consideration by the Ukrainian government.

This chapter provides insights into the concept and components of Ukraine's energy security, principal energy problems and challenges to the Ukrainian energy strategy in new geopolitical reality caused by the conflict with Russia. The chapter discusses the environmental element of energy security, challenges and solutions related to the efficient utilisation of energy resources, opportunities for increasing energy generation from renewable sources, and the implementation of the European Union's environmental rules and standards in the Ukrainian laws and practice.

Energy Security of Ukraine: Critical Issues

Since 1991, issues related to of energy security have been a subject of close attention to the Ukrainian policymakers throughout Ukraine's independence. As a result of the Crimea's annexation and the armed conflict in the east of Ukraine, a number of additional, radically new challenges and threats to ensuring the nation's energy security arose. However, to date, the concept of energy security and what it entails are not yet well embedded in Ukrainian legislation. There are no single universally accepted definitions of energy security and a consolidated approach to the energy policy of Ukraine.

Despite its widespread use, the term *energy security* is defined only in some industry-specific laws and acts of a declarative or advisory nature. One of the definitions describes energy security as the timely, complete and uninterrupted supply of high-quality fuel and energy to consumers, prevention of the harmful influence of the transportation, transformation and consumption of fuel and energy resources on the environment, in the conditions of modern market relations, trends and indicators of the global energy market (CMU 1998). Another Act describes energy security as the conditions in the economy which allow for the efficient use of the country's energy resources, ensuring the availability of a sufficient number of energy producers and suppliers in the energy market, and the availability, variety and environmental friendliness of energy resources (MEDTU 2013).

Certain issues of energy security are also regulated by the Principles of National Security Act (VRU 2003), the Ukraine's President's Acts that validate the decisions of the National Security and Defense Council of Ukraine, regulations of the Cabinet of Ministers and industry-specific legislation. The strengthening of the country's energy security is also marked as a key target of the Energy Strategy of Ukraine up to 2035, adopted by the government in 2017 (CMU 2017a) and that replaced the Energy Strategy of 2013.

The energy security of a nation is a broad category that may be considered through the prism of three major components: (1) the availability of energy resources and services for the sufficient and uninterrupted supply to consumers, (2) energy efficiency, and (3) environmental sustainability. The first component includes the promotion of domestic energy generation, the minimisation of the dependence on imports, the diversification of power supplies, the creation of competitive and transparent energy markets, the availability of strategic reserves and the protection of critical energy infrastructure. Energy use efficiency, environmental safety and the sustainable development of the energy sector, particularly via increasing energy generation from renewables, make a significant contribution to the enhancement of energy security.

Despite acknowledging the high importance of ensuring energy security, during the years since it gained its independence in 1991, Ukraine remained the least energy secure country in the large energy user group consisting of 25 countries. The 2016 edition of the International Index of Energy Security Risk, which aims to assess the likelihood of energy shock of any kind and how it may impact a country's economy, shows Ukraine's energy security risks scores, based on a combination of 29 various metrics. This demonstrates that the risk of disruption to the energy supply in Ukraine is higher in comparison to any other country assessed by the study (Institute for 21st Century Energy 2016).

The energy sector of Ukraine is facing the following major threats:

- high energy intensity and low energy efficiency of Ukraine's economy;
- lack of domestic energy production;
- dependence upon imports of energy resources;
- low diversification of power supplies (i.e. dependence on one supplier);
- lack of strategic reserves for emergencies;
- inadequate protection of strategic energy assets and infrastructure;
- outdated energy infrastructure, used far beyond the term of its safe operation; and
- environmental problems caused by power generation from fossil fuels.

Throughout its independence, Ukraine relied upon energy supplies from other countries. According to the Energy Strategy of Ukraine up to 2035 (CMU 2017a), in 2015, Ukraine depended upon imports, serving 51.6% of its energy needs. Generally, this aligned with an average European level. The problem, however, is that, not only is Ukraine largely dependent on energy imports, it is dependent on one, monopolistic supplier. Prior to 2014, most of its imported oil and gas (approximately 70%) and nuclear fuel (100%) was received from Russia (CMU 2013). Thus, the conflict with Russia had a serious negative impact on the energy security of Ukraine. Overcoming this high dependency on imports of energy resources from Russia is a high-priority task of the current energy policy of Ukraine. To summarise, in the short-term, Ukraine's energy policy aims to prevent problems with the disruptions in the supply of energy. This goal may be achieved through certain elements of energy security, such as greater energy efficiency, increased domestic energy production, the diversification of imports of energy resources, the formation of strategic reserves, the protection of critical energy infrastructure, and forging closer ties with the energy markets of the European Union.

In 2011, to ensure its energy security, Ukraine joined the Energy Community, an international organisation dealing with energy policy, established by an international Treaty in October 2005 in Athens, Greece, between the European Union (EU) and other nations in South East Europe, and the Black Sea region, to extend the EU internal energy market to South East Europe and beyond. Like other contracting parties, Ukraine committed itself to implement a number of relevant European directives and regulations, develop an adequate regulatory framework and liberalise its energy markets in line with the legislation of the European Union (*aquis communitaire*). Latterly, this became part of Ukraine's obligations under the Association Agreement between Ukraine, on the one hand, and the European Union, the European Atomic Energy Community and their member states, on the other, signed in 2014 and, after a long period of ratification by the member states, entered fully into force on the 1st of September 2017.

Environmental compatibility is also an important component of Ukraine's energy security. According to the 2018 Environmental Performance Index (Yale Centre for Environmental Law and Policy 2018), Ukraine is ranked 109 out of 180 countries, having fallen 65 places over the last two years. As a comparison, neighbouring Russia took 52nd place, while Belarus is in 44th position. Thus, the environmental situation in Ukraine requires a far-reaching policy aimed at significant improvement. Ukraine's energy intensity and its greenhouse gas reduction targets are among the worst in Europe. However, due to focusing on the military conflict in its territory and immediate economic concerns, Ukraine's decision-makers have not yet been able to address sufficiently the necessary reforms.

Energy Intensity and Energy Efficiency

Energy intensity, i.e. the amount of energy used per unit of gross domestic product (GDP), in Ukraine is on average two to four times higher than in the UK and countries of Central and Eastern Europe.

As a rule, energy intensity is determined as the ratio between gross inland energy consumption (GIEC) and GDP, calculated for a calendar year. To allow for the comparison and to monitor trends, GDP is calculated in constant prices (i.e. using prices of a fixed base year) to exclude the impact of inflation, and based on purchasing power parity (PPP), that is the rate of currency conversion which equalises the purchasing power of different currencies, by eliminating the differences in price levels between countries. For these purposes, GDP is measured in *international dollars* (also known as the *Geary-Khamis dollar*), a currency unit used by economists and international organisations to compare the values of different currencies, while GIEC is measured in 1000 tonnes of oil equivalent (ktoe). Table 6.1 demonstrates the level of energy intensity of Ukraine compared to other countries.

Ukraine's high energy intensity is determined by several factors. A considerable portion of GPD is created in the resource and energy intense sectors which include: the steel industry, mining, cement and chemical industry and machine-building industry. A high degree of wear of fixed assets, dated technology, low energy tariffs, that for a long time were artificially kept below market prices, and high costs of energy production are the principal impediments to the modernisation of energy assets. In addition, the high level of energy intensity stems from a considerable waste of energy in the course of power generation, transportation and distribution of electricity and heating energy, as well as the inefficient utilisation of energy by households. At the end of 2016 the degree of wear of fixed assets in the extraction industry was 54.6%, while in the field of transportation of electricity and natural gas—it was 62.1% (Ukrstat 2017a).

The challenges related to raising energy use efficiency are at the forefront of the energy policy of Ukraine. The principal state planning acts in the field of energy efficiency are the State Special Purpose Economic

Table 6.1 Energy intensity at purchasing power parity (PPP) of the economy of Ukraine, United Kingdom and selected European countries, 2016

Indicator	Germany	France	Italy	Poland	Lithuania	United Kingdom	Ukraine
Gross inland energy consumption, ktoe	317,268.3	248,745.5	154,748.0	99,929.6	7034.0	189,439.8	91,312.6
GDP, PPP (in constant 2011 international USD), bln	3658.9	2546	2101.1	988.6	80.4	2578.5	327.2
Energy intensity, gross inland consumption/GDP (ktoe/bln 2011 international USD)	86.7	97.7	73.7	101.1	87.5	73.5	279.1

Source Compiled by the author from Eurostat's and World Bank's data (Eurostat 2018; World Bank 2018)

Programme for Energy Efficiency and Development of the Sphere of Production of Energy Products from Renewable Energy Sources and Alternative Fuels for 2010–2017 (CMU 2010) and the National Energy Efficiency Action Plan for the Period until 2020 (CMU 2015).

The Energy Efficiency Action Plan set a target to, by 2020, reduce final energy consumption to 9% of the annual average amount of final inland energy consumption for 2005–2009 (CMU 2015). It foresees a number of measures related to the increase of energy efficiency in private households, industry, transport and services sectors, including investments in heat insulation of buildings; state financial support for raising energy efficiency of residential apartment buildings and houses; implementation of energy consumption metering across the board (for all business and individual customers); the improvement of construction standards and the introduction of minimum standards for industrial equipment; launch of energy audit and certification; engagement of energy service companies; and energy labelling.

Energy wastage, as related to heating private households, amounts to 60%, or 3 billion US dollars per year (CMU 2016a). Since October 2014, the Ukrainian government has been implementing a state programme to partially compensate individuals and homeowner associations for the cost of loans for undertaking energy efficiency measures (the 'warm loans' programme). The loans aim to facilitate the purchase of alternative fuel boilers to reduce natural gas and electricity consumption; ensure better insulation of buildings; finance installation of water and heat metres; and provide funding for replacing windows, the modernisation of lighting etc. From the launch of the 'warm loans' programme to December 2017, about 400,000 families took advantage of these opportunities, and more than UAH (Ukrainian hryvna) 6 billion of investment was made in the economy (SAEE 2017a). The state budget's expenses, targeted to compensate part of the cost of the loans for the purchase of energy efficient materials and equipment, amounted to more than UAH 1.8 billion. In addition, more than 150 local programmes provided funding to complement 'warm loans'. On average, the savings of natural gas after thermal modernisation of a detached house amount to 29% (SAEE 2016). As a result, for the time from 2014 to September 2017, the State Agency of Energy Efficiency and Energy Saving of Ukraine reported that the total saving of energy resources amounted to 147 million cubic metres (of gas equivalent) (SAEE 2017b).

The next step in financing energy efficiency measures is to establish the Energy Efficiency Fund for the financial support of energy modernisation. By doing this, Ukraine would fulfil its international obligations in the field of energy efficiency, including Directive 2012/27/EU of the European Parliament and of the EU Council, following ratification of the Treaty establishing the Energy Community. The Energy Efficiency Fund Act (VRU 2017a) was adopted by the Verkhovna Rada (Parliament) of Ukraine on 8 June 2017. The Act stipulates that the Energy Efficiency Fund is a state-run company that should facilitate energy saving at the level of final energy consumption. The fund will be partially financed by the state budget of Ukraine and will also be able to attract grants and other financing from the government, various agencies, and foreign institutions. The fund will provide grants and partial compensation of costs for the implementation of energy efficiency projects to legal entities and individuals and will carry out the financial assessment of risks of such projects. In addition, the fund will monitor and appraise a range of projects aimed at increasing energy efficiency. On 20 December 2017, the Cabinet of Ministers of Ukraine established the Energy Efficiency Fund and adopted the procedure for the use of budget funds for the work of this organisation (CMU 2017b, c), which is a significant and positive step towards the creation of an energy efficient society.

Supply of Energy Resources

The conflict with Russia forced Ukraine to seek ways of reducing dependence on Russia for the supply of energy resources. In relation to the supply of energy resources, Ukraine is facing three key and challenging tasks: (i) to support domestic energy producers in order to increase the share of own energy resources, (ii) to diversify imports of energy resources, and (iii) to improve the structure of total primary energy supply (TPES). The latter refers to the total amount of primary energy that a country has at its disposal, including energy extracted from natural resources, imported energy minus exported energy. TPES could be improved by lowering the share of natural gas in it, through the development of renewable energy sources and by increasing the share of green energy in the energy mix of the country.

Ukraine's economy has traditionally experienced power shortages. Due to the annexation of the Crimea and the military conflict in the east of Ukraine, as well as the economic recession, this problem became more acute in 2015. For instance, coal mining decreased by 36.7% compared to 2014 (BP Statistical Review of World Energy 2016). At the same time, Ukraine ranks third in Europe, after Norway and the Netherlands, in terms of proved gas reserves (BP Statistical Review of World Energy 2017). However, many gas and oil reserve sites are undeveloped, or the volume of extraction is low. By increasing the efficiency of energy consumption and the production volumes, Ukraine has the opportunity to completely eliminate the need to import gas (CMU 2016b).

However, the increase of energy production requires constant and significant investment in new technology, the development of new fields and intensification of production in the developed reserves. While large deposits often appear to be considerably exhausted, smaller deposits lie at a great depth and are more difficult to explore, which requires investment in research and new, often expensive technology. The domestic public and private financing is limited, while any possibility of an inflow of foreign investment requires further improvements to the business climate in Ukraine and a need to solve a range of sectoral problems. These include:

- unstable fiscal policy (in particular, significant increase of rent payments for subsoil use in 2014–2015);
- discrimination in tax rates for private companies compared to state-owned enterprises;
- poorly designed tariffs that were historically capped below market prices;
- high level of state monopoly over the production and supply of energy resources (i.e. for many years state-owned companies enjoyed a monopolistic position over exploration, production, export and

import operations and benefited from privileged rights to obtain new licences without an auction) (Borzhemska 2017);
- overregulation of business;
- barriers to access land resources and subsoil; and
- high cost and lengthy time required for connecting a new energy facility to the grid.

To foster the extraction of energy resources, in 2016–2017 Parliament decreased the rent payments for subsoil use by 50%, which applied to private extracting companies. Positive changes were also observed regarding access to land plots and subsoil for exploration and commercial development of oil and gas fields: on 1 March 2018, the Parliament adopted the Act on Amending Certain Legislative Acts of Ukraine Concerning Deregulation in the Oil and Gas Sector. The Act simplified the procedure for launching exploratory and commercial development of oil and gas fields, eased access to and use of land plots and canceled unnecessary duplicating permits (VRU 2017e).

Ukraine also has one of the greatest potentials in Europe for producing energy from renewable sources of almost all energy types (wind, solar, hydropower, biomass, biofuel, geothermal). With a combination of abundant resource potential and state-supported renewable energy promotion schemes (feed-in tariff scheme), Ukraine is a very promising renewable energy market (OECD 2012).

In October 2014, as part of its commitments to the Energy Community, Ukraine adopted the National Renewable Energy Action Plan (NREAP) until 2020 (CMU 2014). In this Plan, renewables potential of the nation is assessed at 68.6 million tons of oil equivalent per year. Therefore, renewable energy could meet about 50% of Ukraine's energy needs.

The NREAP aims to achieve an 11% renewable energy share in Ukraine's total final energy consumption by 2020. Total final consumption for a country is the aggregate of all energy that is used by consumers to serve their energy needs. The share of 11% in the total final energy consumption constitutes approximately 8% in the country's TPES, an aggregate of all energy going into the energy sector. Due to economic recession, achieving this target presents a challenging task.

Thus, in 2016, the share of renewable sources of energy in the TPES amounted to just 3.9% (approx. 5.8% of total final energy consumption) (Ukrstat 2017b). Despite little progress made since the adoption of the NREAP in 2014, the Energy Strategy has set even higher targets, to be reached by 2035: to achieve a 12% share of renewables in the TPES by 2025 and 25% by 2035 (CMU 2017a).

Visible progress should be noted regarding the diversification of imports of energy resources. An ambitious plan to receive no more than 30% of supplies from one source for each type of energy resources is in place. Within Ukraine's course on the Europeanisation of local energy markets, it is expected that such diversification will be feasible, by adopting the European Union's technical and legal standards, forging closer ties with the European energy markets, and through the creation of competitive national energy markets, that are currently in the spotlight of the Ukrainian energy reforms.

More specifically, Ukraine has managed to engineer reverse-flow supplies of natural gas from Europe and fully abandon the need to import gas from Russia, which began at the end of 2015 until the beginning of 2017. The recent successful reform of the Ukrainian gas market created opportunities for the direct supply of natural gas by foreign gas traders. The gas supply contract with Norwegian Statoil, which was signed in 2014, serves as additional security tool aimed at the expansion of geography of supplies and the number of gas suppliers. In order to further diversify the sources of supply of nuclear fuel, in 2014, Ukraine extended the contract with Westinghouse Electric Sweden, a subsidiary of Westinghouse Electric Company (USA), to until 2020. Currently the most pressing task is to reduce the dependence on the supply of anthracite coal from Russia and the territories in the east of Ukraine, where the conflict continues. In part, the problem might be resolved by foregoing anthracite coal in favour of gas coal (the type of coal that is rich in volatile hydrocarbons, making it a suitable source of domestic gas)—this requires retrofitting thermal electric power stations and thermal heating plants. Thus, in 2017, Ukraine cut consumption of anthracite coal by 46%, or 4.1 million tons. This resulted in the savings of 2.3 million tons of anthracite coal due to the increase in power generation by nuclear plants and hydroelectric power plants, and 1.8 million tons of anthracite was replaced by gas coal produced in Ukraine (Interfax 2018).

Strategic Reserves

Ukraine's energy crisis, caused by the deterioration of relations with Russia and economic recession proved that the creation of strategic reserves of energy resources for emergencies is one of the conditions of the nation's energy security. Insufficient domestic production of fossil fuels, dependence on imports, limited capacity to form adequate supply of energy resources, the threat of the reduction or interruption in the supply of power, as well as requirements set in the EU directives, explain the urgent need to create minimum reserves of fossil fuels.

The EU Council Directive 2009/119/EC, of 14 September 2009, imposed an obligation on EU member states (and on Ukraine, due to its membership in the Energy Community and owing to entering into the Association Agreement with the European Union) to maintain minimum reserves of crude oil and/or petroleum products. To implement this directive, Ukraine had to adopt relevant laws, regulations and administrative procedures and create, by the end of 2022, total oil reserves maintained at all times, corresponding to 90 days of average daily net imports or 61 days of average daily inland consumption, whichever is greater. On 8 April 2015, the Cabinet of Ministers of Ukraine approved the Plan for Implementation of Directive 2009/119/EC, developed by the Ministry of Energy and Coal Industry of Ukraine. However, no genuine progress could be reported with respect to the implementation of the plan: despite the set deadline of December 2016, the Ministry of Energy and Coal Industry has been failing in the preparation of the draft Act on Maintenance of Minimum Reserves of Crude Oil and Petroleum Products. The plan developed by the Ministry was replaced by the List of Measures for Fulfilment of the Association Agreement between Ukraine, on the one hand, and the European Union, the European Atomic Energy Community and their member states, on the other hand. The document was approved by the Ukrainian Government on 25 October 2017 and entered into force on 17 March 2018 (CMU 2017d). Now the State Reserve Agency of Ukraine, a central governmental authority that implements policy in the field of state material reserves, took the lead on preparation of the new model of functioning and financing of minimum reserves of crude oil and petroleum products and the respective draft Act. The model provides for creation of an independent stockholding agency, fully responsible for meeting Ukraine's minimum reserve requirements. According to the State Reserve Agency, the structure of minimum stock will consist of 30% of crude oil and 70% of petroleum products (SRA 2017).

There is also an opinion that Ukraine needs strategic reserves of natural gas and coal (Unigovskiy 2016) considering the share of these energy resources in the TPES: 27.9 and 32.4%, respectively, in 2016 (Ukrstat 2017a). Thus, Ukraine has the most powerful network of underground gas storage facilities in Europe. The total active capacity of Ukraine's underground gas storage facilities located in controlled territories is over 30 billion cubic metres, or almost one third of the EU's gas storage facilities. As of 2018, the obligation to create reserve stock of natural gas in the amount of up to 10% of the contemplated monthly supply applies to gas suppliers only. For 2016–2018 the Government decided that the reserve amount should be zero, and only in case of an emergency situation, it should amount to 10%. However, the underground gas storage facilities may be used not only to cover seasonal spikes in consumption and to store gas of European and Ukrainian supplies, but also to create strategic reserves of the state. This requires reconsideration of Ukraine's energy strategy in this area and the development of the necessary legal framework.

Protection of Critical Energy Infrastructure

Energy systems and assets are traditionally considered to be critical infrastructure, vital for the proper functioning of a nation. For Ukraine, the protection of critical energy infrastructure from an accident or attack became more important than ever due to the conflict with Russia. Apart from seizure of the energy infrastructure in the Crimea and Black Sea shelf, the authorities disclosed in the media certain incidents, such as seizure of the gas distribution station of State Joint Stock Company Chornomornaftogaz in Kherson oblast by the armed unit of presumably Russian army in March 2014 (GPO 2014); explosions in the gas distribution hub on the Urengoy—Pomary—Uzhgorod gas pipeline in

May and June 2014 as the damage to the pipes was caused by unauthorised persons (Isachenko 2014; NPU 2014); seizure of the control station on the main gas pipelines in Kramatorsk (Poltava oblast) in May 2014; threat of terroristic attacks at nuclear power objects (e.g. the attempt of armed people to seize the Zaporizhzhia Nuclear Power Station in May 2014) (Sukhodolia 2014); and cyberattacks on energy facilities (e.g. cyberattack on the power grid in December 2015 that led to the disruption of electricity supply to end consumers; ransomware virus 'Petya' that hit Ukrainian governmental agencies, banks and businesses, including energy companies, and then spread all over the world in June 2017).

Protection of the energy infrastructure vital for the normal functioning of the economy has come into focus for the government just recently and the country has not yet defined what critical energy infrastructure includes. The existing legislation is mainly concerned with the protection of individuals from the aftermath of technological accidents, rather than the prevention of subversive actions. The issues of the protection of energy infrastructure are resolved at the industry and government departments' levels, while a lack of proper coordination between civil protection, terrorism prevention and other governmental authorities should be noted.

On 29 December 2016, the Resolution of the National Security and Defense Council of Ukraine (NSDCU 2016) enacted by the President's Decree no. 8/2017, dated 17 January 2017, ordered the relevant governmental agencies to draft a concept for the creation of a critical infrastructure protection system and to develop a draft law on critical infrastructure and its protection. Despite the fixed two-months deadline, the draft law has yet to be submitted to Parliament.

At the same time, certain positive steps were taken in strengthening Ukraine's cybersecurity: on 5 October 2017, the Verkhovna Rada of Ukraine adopted the Act on Basic Principles of Ensuring Cybersecurity of Ukraine (VRU 2017d). The Act introduced a number of concepts such as: cybersecurity, cyber-protection, cyberattack, cyber-threat, cyberspace, cyberterrorism, objects of critical infrastructure, and objects of critical informational infrastructure, among others. According to this law, enterprises and institutions carrying out activities in the energy sector can be included in the list of objects of critical infrastructure.

The procedure for the formation of such a list, requirements for cyber-protection of assets included in the list and an audit of their informational security are to be further approved by the Cabinet of Ministers of Ukraine.

The Act on Basic Principles of Ensuring Cybersecurity of Ukraine provides for creation of the national cybersecurity system as the aggregate of authorities providing cybersecurity, including the State Service of Special Communication and Information Protection, National Police, Security Service of Ukraine, Ministry of Defense, General Staff of the Armed Forces, intelligence agencies, and the National Bank of Ukraine, and sets their functions. The law also regulates the creation and functioning of the National Telecommunications Network and sets the tasks for the government response team, entitled CERT-UA, in relation to computer accidents. The coordinating role in the field of cybersecurity will be taken by the President of Ukraine through the National Security and Defense Council and its National Coordination Center of Cybersecurity. The Government has yet to develop all necessary by-laws to ensure implementation of the Act.

Environmental Sustainability

Despite Ukraine is a party, since 1997, to the United Nations Framework Convention on Climate Change, which deals with mitigation of greenhouse gas emission, signing a number of other international environmental treaties, declares environmental safety and sustainable development as a national priority, and has formed a number of national and international environmental programmes, until recently, environmental protection and climate change were hardly present on Ukraine's political agenda.

The scope of the Ukrainian environmental legislation is broad and comprehensive (more than 300 legal acts). However, the environmental legislation is largely declaratory in nature and does not include all of the essential enforcement mechanisms for the implementation of legal acts and international agreements. Many of the acts are not harmonized with each other, and legislation undergoes limited analysis of its impact and is frequently changed. Despite the decentralisation reform launched in 2010–2012, environmental protection is still strongly centralised at the national level, leaving little space for local initiatives and providing no well-functioning mechanism for coordination. Finally, underfunding and misappropriation of public funds, lack of monitoring and evaluation mechanisms, and poor access to information on the progress and impact of national and international environmental programmes contribute to the environmental problems that Ukraine currently faces. In addition to the much-publicised Chornobyl disaster, the principal problems include air pollution; poor quality of water resources; land degradation; problems related to solid and hazardous waste utilisation; biodiversity loss; human health issues associated with environmental risk factors; and climate change (IBRD and World Bank 2016).

The recession, conflict with Russia and loss of control over considerable parts of Donetsk and Luhansk oblasts only made things worse, as the main goal of all recent energy reforms was to save money and end the dependence on gas imports from Russia.

According to the 2016 edition of the International Index of Energy Security Risk (Institute for 21st Century Energy 2016) Ukraine occupies the worst position among 25 countries under CO_2 intensity of GDP. Notwithstanding the country's general downward trend regarding the volume of carbon emissions and emissions intensity (for example, carbon emissions intensity of Ukraine fell gradually from 1.66 kg per 1000 dollars of GDP in 1997 to 0.71 kg per 1000 dollars of GDP in 2016 [WDA 2018]), the level of emissions in Ukraine is still extremely high when compared to its relatively low GDP. This is explained primarily by the high consumption of coal, high energy intensity of the national economy and lack of modern, environmentally-friendly technologies.

Environmental Impact of the Conflict in Donbas

The international community is also concerned with the environmental impact of the military conflict in Ukraine's highly industrialised Donbas region famous for its high concentration of coal mining, chemical and metallurgical industries. The conflict not only resulted in a number of civilian health risks, but also in the potentially long-term damage to the environment. Damage to industrial facilities, coal mines, water supply and other infrastructure in certain cases led to the accidental release of pollutants and degradation of the quality of water. Among dozens of facilities damaged by shelling are the Zasyadko coal mine, a storage of chemicals at Yasynivskyi coke and chemical plant, the chemical plant in Makyivka, the Lysychyansk oil refinery, an explosives factory at Petrovske, and an oil storage facility at Slavyansk thermal power plant. As an example of the scope of the damage, the Zasyadko coal mine in Donetsk used to produce 4 million tons of coal annually and was one of the region's economic flagships. A release and explosion of methane in March 2015, due to the heavy shelling of a nearby airport in Donetsk, killed 33 of the 200 miners underground at the time (Zoï 2015).

According to the statement of the former Minister of Ecology and Natural Resources Andriy Mokhnik made in 2014, two-thirds of Donbas coal mines have been flooded as a consequence of military actions, and this means that the contamination of underground waters and potable water in the entire coal basin has occurred (Solonyna 2014).

Bellingcat, an investigative search network founded by the British network activist Eliot Higgins, performed its own independent study of the impact of the military conflict on the region's ecology, based on open source data. Analysts of the agency have confirmed that there is a high probability of irreversible consequences, not only for the ecology of Donbas, but also for the neighbouring regions (Roberts 2017). According to Bellingcat, the list of the most hazardous enterprises includes the Avdiivka coke and chemical plant, which suffered from more than 300 shells explosions, Novgorodske phenol plant, water purification stations near Donetsk, Mykhailivka Electrical Transformer Station, Nikitovskyi mercury mine plant, and Luhansk thermal power plant (Zwijnenburg 2017). It is hardly an exaggeration to say that these facilities should be viewed as extremely high-risk. Ukraine, together with the international community, has yet to find ways to document, assess and address the damage caused, as well as to prevent its aggravation.

Greenhouse Gas Emissions Goals

Notwithstanding the above, certain progress may be noted in the environmental field. On 22 April 2016, together with more than 150 countries, Ukraine signed a new climate treaty, the Paris Agreement, that replaces the Kyoto Protocol, and undertook a commitment to keep greenhouse gas emissions at 60% of the 1990 levels. For a comparison, pursuant to Annex B to the Doha Amendment to the Kyoto Protocol, Ukraine's allowed greenhouse gas emissions for 2013–2020 were equal to 76% of the 1990 level (United Nations Climate Change 2012). In fact, a 60% level has already been reached by Ukraine: in 2012 the greenhouse gas emissions amounted to 42.6% of 1990 level (INDC 2016). However, presently, Ukraine also has to take into consideration the necessity to reconstruct ruined industrial facilities and infrastructure in the Donbas region after assumed restoration of its territorial integrity and state sovereignty. Due to this, the Ukraine's Intended Nationally-Determined Contribution is based on the eventual increase of the production of metal, non-metal construction items, and other products required for such restoration.

While the demand for energy is increasing, the reduction of greenhouse gas emissions may be achieved by structural changes in the economy, enhancing energy efficiency, changing the country's energy mix towards less carbon-intensive fuels (e.g. low-carbon gas instead of carbon-intensive coal); and the deployment of renewable energy facilities.

Renewable Energy Development

Installed capacity of renewable energy sources in Ukraine tends to grow annually, which increases the share of renewables in the TPES. An exception to this was 2014, which showed a decrease due to the loss of renewable energy assets in the Crimea and the eastern territories of Ukraine, as well as a sharp decline in investment caused by the economic and political crisis. Starting from 2009, the average growth rate of installed capacity of renewable energy sources has amounted to 31%. As of January 2017, the capacity of renewable energy facilities in Ukraine working under the feed-in tariff was 1117.7 MW. By the end of 2016, the renewable energy sector was comprised of 170 companies and 291 energy facilities. In 2016, 120.6 MW of capacities were set into operation. The maximum growth was shown by solar energy: 36 new producers and 47 new facilities (Energy Efficiency Secretariat 2017). In 2017, the total capacity of renewable energy facilities in Ukraine increased by 23% and at the end of the year, it amounted to 1375 MW produced by 376 facilities (SAEE 2018).

This, however, is still insufficient to achieve the target of 11% of renewables in the TPES by 2020 (CMU 2014). More active state support is required in the development of renewable energy facilities. Since 2009, Ukraine has been making efforts to financially encourage power generation from alternative sources. Such encouragement includes the introduction of the 'green tariff' policy, open until 2030, which is a feed-in tariff scheme (i.e. the guaranteed obligation of the state to purchase green energy generated by domestic producers), and the introduction of certain tax benefits for producers of alternative energy. Some examples of tax benefits include an exemption from VAT on imported goods and exemption from customs duties on transactions involving the import of equipment for renewable energy generation, energy saving equipment and materials, means of measuring, control and management of energy resources, provided that such equipment and materials are used by a taxpayer for own production needs and no identical goods of the same quality are produced in Ukraine (Dukunskyy and Zharikov 2016; VRU 2010).

The rates of the feed-in tariff, the scope and requirements for its use have been changing during the period of its existence. For several years, feed-in tariffs in Ukraine for electricity produced by ground-mounted solar power plants were the highest in the world, but this was not always economically justified. In June 2015, significant changes were introduced to the procedure for the calculation of the feed-in tariff and where it can be used. According to these changes, the feed-in tariff applies to electricity produced by solar power plants and wind farms (including solar and wind installations of private households) and small hydropower plants, as well as electricity generated from biomass, biogas and geothermal energy. The feed-in tariff is tied to the EUR/UAH exchange rate and is adjusted on a quarterly basis. The tariff depends on the date when the energy facility is set into operation. A premium is paid on the tariff depending on the share of local content (share of components of Ukrainian origin used during the construction of the renewable energy facility). The energy facilities using more than 30% of components produced locally will receive an additional premium, in the amount of 5% paid on top of the regular feed-in tariff, while using more than 50% gives a 10% increase. As of March 2018, 17 Ukrainian companies were benefiting from the premium to green tariff for the use of equipment of the Ukrainian origin.

The Electricity Market Act of Ukraine adopted in 2017 (VRU 2017b) and subordinate legislation introduced additional positive changes allowing producers and potential producers of green energy to receive financing: it became possible to enter into a power purchase agreement (PPA) with a guaranteed purchaser before commencing construction or putting into operation a renewable energy facility (pre-PPA concept). In addition, PPAs now may be concluded for the entire period of validity of the feed-in tariff scheme (contrary to annual contracts as was previously the case), which adds certainty to the investors. As a result of these changes, in December 2017, the Overseas Private Investment Corporation (OPIC), the US government's development finance institution, approved USD 400 million for financing and political risk insurance to support construction by EuroCape Ukraine I LLC (a Ukraine-based subsidiary of an international renewable wind energy group) of the largest wind farm in Ukraine, with a 500 MW capacity. The project will strengthen energy security in Ukraine by increasing Ukraine's wind generation capacity by 40% (OPIC 2017; NEURCU 2017).

In terms of the next steps of state support of renewable energy in Ukraine, the removal of regulatory barriers related to access to land plots by green energy producers (overregulated, non-transparent and lengthy land allocation procedures) needs to take place and simplifying the procedure for obtaining permits and approvals for commencement of construction and ensuring timely connection to the grid of the constructed renewable energy facilities are also of key importance. The development of the renewable energy may also be stimulated by the introduction of additional tax and customs benefits for green energy producers.

For a long period, the renewable energy market was monopolised by Ukrainian oligarchs who persuaded the government to set up barriers to entry onto the market by other investors. The two largest industrial groups in the field (Activ Solar and DTEK) have links with influential Ukrainian billionaires Sergiy and Andriy Kliuiev and Rinat Akhmetov (Nowak 2015). However, since the cancellation of the local content requirement in 2015 and further liberalisation of the Ukrainian renewable energy market, the latter became more attractive for large international players and small and medium-sized businesses in Ukraine. As of today, the largest solar power operator is CNBM (China) which in 2016 acquired 10 solar power farms previously controlled by Activ Solar. In the wind power sector, the key players are DTEK Wind Power and Windparks of Ukraine which are controlled by Ukrainian large industrial groups (DTEK 2017).

Solar energy is the most rapidly developing sector of Ukraine's renewable energy market. Solar projects have become increasingly popular not only among large-scale business, but also among small and medium-sized enterprises, for example, among companies and investors whose primary activity is agri-business. In addition to ensuring uninterrupted energy supply, the farmers can gain profits from selling the excess electricity at the feed-in tariff. The 2015 changes in the green tariff scheme also encouraged the installation of solar panels on residential buildings. This trend has been growing continuously since 2016.

The wind energy sector is growing more slowly than the solar energy sector. The explanation for this is that wind projects require larger investment and take a longer amount of time to launch. Typically, wind project development requires between three and five years from the time of project initiation to project commissioning. Wind parks are more complicated to instal and require special maintenance. Furthermore, the wind power sector is more regulated than the solar energy sector. That is why this market is dominated by large companies while smaller investors are underrepresented (Baker Tilly 2018). The generation of power from biomass is very popular with heat supply companies (both public and private), industrial enterprises, as well as other small and medium businesses and individuals using generated energy for their own consumption.

In terms of potential for increasing energy generation from renewables, the most attractive regions in Ukraine are southern and south eastern oblasts. Crimea annexed by Russia and the steppe zone of Ukraine (including Odesa, Mykolaiv, Kherson oblasts, and forest-steppe Vinnytsia oblast) have the highest concentration of solar energy farms, due to a significant level of solar irradiation in those regions. The greatest wind resources in Ukraine are in the Carpathians, southern coast of Ukraine adjacent to the Black and Azov seas, Donbas region, and windy areas along the Dnipro river in central Ukraine. The production of energy from biomass received a strong boost in the agricultural and forestry parts of Ukraine, while small-scale hydropower projects are linked to the location of small rivers, mainly in Vinnytsia, Kirovohrad, and Ternopil oblasts. The western part of Ukraine (in particular, Zakarpatska and Chernivetska oblasts) has the largest potential for small hydro energy projects (SAEE 2017c).

One of the new opportunities for Ukraine's renewable energy could be the use of the site close to the Chornobyl nuclear power plant (which in 1986, suffered from one of the world's worst nuclear disasters) for construction of solar power farms. In addition to providing the country with cheap, clean energy, the project dubbed *Chornobyl Solar* can make Ukraine less dependent on supply of energy resources from Russia. During the first stage, a number of land plots of 2500 ha have been offered for deployment of solar panels with a total installed capacity of approximately 1.2 GW (SAEZM 2016). In the future the project may be expanded. To make this project possible, Ukraine introduced important legislative changes to allow leasing land plots located in the exclusion zone (an area surrounding the Chernobyl nuclear power plant where radioactive contamination from the disaster is at its highest and where public access and inhabitation are restricted) and established low land lease rates. More than 50 domestic and foreign investors (including Chinese, Danish, French, United States, German, Belorussian companies) have expressed interest in the development of solar power projects in the Chornobyl exclusion zone.

Implementation of European Environmental Assessment Standards

After it became a member of the Energy Community in 2011, Ukraine committed to harmonise its environmental assessment procedures and to adopt laws on environmental impact assessment and strategic environmental assessment, which later became a part of its obligations under the Association Agreement between Ukraine and the European Union. After numerous debates related to the attempt to find a balance between the implementation of EU relevant practices and creation of additional, burdensome, regulatory barriers for business, a veto of the President and subsequent revisions, on 23 May 2017, the Ukrainian Parliament adopted the Environmental Impact Assessment Act (VRU 2017c). The Act provides for implementation of the EU model for evaluation of the impact of potentially harmful projects in public and private sectors on the environment, to prevent damage to people's lives and mitigate health and environmental degradation. It establishes an obligation to obtain in advance a positive expert opinion with regard to the impact on the environment of planned projects for a broad range of business activities, depending on their scale, including those in the energy sector, metallurgical and chemistry industries, construction, waste management, extraction industry, agriculture and forestry, food processing industry, consumer goods manufacturing, infrastructure projects, and tourism and recreation.

On 20 March 2018, the Verkhovna Rada of Ukraine also adopted the Strategic Environmental Assessment Act (VRU 2018) which provides for the creation of a mechanism that would assess the impact of state programmes and policies on the environment and health.

Thus, Ukraine made certain progress in the implementation of its international obligations in the field of environmental assessment. The adoption of the mentioned two acts is undoubtedly a positive step, that is intended to predict environmental problems of public and private projects and develop approaches to how to prevent or mitigate them. Nevertheless, serious efforts are still to be made to ensure the practical implementation of environmental impact assessment and strategic environmental assessment procedures.

Conclusion

Energy security is still one of the areas where Ukraine is vulnerable. To be successful, the energy security policy must be multi-focused. In particular, it should simultaneously focus on:

- ensuring the uninterrupted availability of energy resources and services on the market, affordable for all consumers;
- creation of competitive and transparent energy markets;
- energy saving and energy efficiency;
- protection of critical energy assets; and
- environmental sustainability of the energy sector.

However, after facing external aggression and internal armed conflict, Ukraine's primary security goals are to reduce the dependence upon power supplies from Russia through the increase of energy efficiency, diversification of supply and 'Europeanisation' of energy markets (i.e. development of competitive, transparent and nondiscriminatory energy markets, under rules and standards of the European Union, as envisaged by the Ukraine–European Union Association Agreement). Due to continuing reforms in the oil and gas sector, electricity sector and in the field of energy efficiency and saving, Ukraine has shown significant progress in reaching these goals. At the same time, insufficient efforts are being made in terms of attracting investments to increase the domestic energy production and create strategic reserves of energy resources. The protection of critical energy infrastructure, which is vital for the normal functioning of the economy, was also out of focus of the government until recently.

The principles of environmental safety have been neglected by the Ukrainian energy policy-makers for a long time. As a result, Ukraine has the most carbon intense economy in Europe and often fails to fulfil its international commitments. Currently Ukraine is on the verge of significant changes—shifting focus towards the environmental component of energy security. International cooperation could make a significant contribution to this field, since most challenges facing Ukraine in this way are common to all nations and experience of states more successful in protecting the environment may be invaluable. Ukraine has chosen the European model of environmental security and is making efforts to implement environmental *aquis communitaire* (legislation of the European Union) into the Ukrainian legal system. In parallel, noticeable progress has been made in the development of Ukraine's renewable energy potential. Further progress requires that sustainable development of the energy sector per the principles of environmental safety is made a key priority of the national energy policy.

Several important lessons may be learned from the analysis of Ukraine's energy security. Ukraine's experience shows that energy security is a part of national security and that dependence on imports of energy products from one country can pose a major threat. Furthermore, analysis demonstrates that energy dependence and other challenges facing Ukraine's energy sector might be overcome by means of successive domestic reforms and international cooperation aimed at 'Europeanisation' of Ukraine's energy markets. Despite the Crimea's annexation and conflict in the east of the country, Ukraine made genuine progress in implementing energy reforms, thus, strengthening its energy security. Although there is still more work to be done, Ukraine is moving in the right direction.

References

Baker Tilly. (2018). *Ukrainian alternative energy sector: Slow but steady development*. http://www.bakertilly.ua/en/news/id1394. Accessed 23 May 2018.

Borzhemska, I. (2017). *Energy sector reform: Ukrainian energy market opens up for investment*. http://bunews.com.ua/investment/item/energy-sector-reform-ukrainian-energy-market-opens-up-for-investment. Accessed 23 May 2018.

BP Statistical Review of World Energy. (2016). https://www.bp.com/content/dam/bp/pdf/energy-economics/statistical-review-2016/bp-statistical-review-of-world-energy-2016-full-report.pdf. Accessed 23 May 2018.

BP Statistical Review of World Energy. (2017). http://www.bp.com/content/dam/bp/en/corporate/pdf/energy-economics/statistical-review-2017/bp-statistical-review-of-world-energy-2017-full-report.pdf. Accessed 23 May 2018.

Cabinet of Ministers of Ukraine (CMU). (1998). *Pro Kontseptsiiu diialnosti organiv vykonavchoi vlady u zabezpechenni energetychnoi bezpeky* (in Ukrainian) [On concept of activity of executive bodies in ensuring energy security of Ukraine]. http://zakon0.rada.gov.ua/laws/show/48-98-%D0%BF. Accessed 23 May 2018.

Cabinet of Ministers of Ukraine (CMU). (2010). *Derzhavna Tsiliova Economichna Programa z* Energoefektyvnosti *ta Rozvytku Sfery Vyrobnytstva Energonosiiv z Vidnovliuvanyh Dzherel Energii ta Alternayvnyh Vudiv Palyva na 2010–2017 Roky (in Ukrainian)* [State special-purpose economic program of energy efficiency and development of the sphere of production of energy products from renewable energy sources and alternative fuels for 2010–2017]. http://zakon3.rada.gov.ua/laws/show/243-2010-%D0%BF. Accessed 23 May 2018.

Cabinet of Ministers of Ukraine (CMU). (2013). *Energetychna Strategiia Ukrainy na Period do 2030 Roku* (in Ukrainian) [Energy strategy of Ukraine for the period until 2030]. http://zakon2.rada.gov.ua/laws/show/n0002120-13. Accessed 23 May 2018.

Cabinet of Ministers of Ukraine (CMU). (2014). *Natsionalnyi plan dii z vidnovliuvanoi energetyky na period do 2020 roku* (in Ukrainian) [National renewable energy action plan until 2020]. http://zakon3.rada.gov.ua/laws/show/902-2014-%D1%80. Accessed 23 May 2018.

Cabinet of Ministers of Ukraine (CMU). (2015). *Natsionalnyi plan dii v energoefectyvnosti na period do 2020 roku* (in Ukrainian) [National energy efficiency action plan for the period until 2020]. http://odz.gov.ua/lean_pro/normdocs/files/NEEAP_1228.pdf. Accessed 23 May 2018.

Cabinet of Ministers of Ukraine (CMU). (2016a). *Kontseptsiia vprovadzhennia mekhanizmiv stabilnogo funktsionuvannia zakhodiv z energoefektyvnosti (stvorennia Fondu Energoefektyvnosti)* (in Ukrainian) [Concept of implementation of mechanisms of stable financing of energy efficiency measures (Establishment of energy efficiency fund)]. http://zakon3.rada.gov.ua/laws/show/489-2016-%D1%80. Accessed 23 May 2018.

Cabinet of Ministers of Ukraine (CMU). (2017a). *Energetychna strategiia Ukrainy do 2035 roku: bezpeka, energoefectyvnist, konkurentospromozhnist* (in Ukrainian) [Energy strategy of Ukraine up to 2035: Security, energy efficiency, competitiveness]. http://zakon2.rada.gov.ua/laws/show/605-2017-%D1%80. Accessed 23 May 2018.

Cabinet of Ministers of Ukraine (CMU). (2017b). *Pro utvorennia derzhavnoi ustanovy "Fond energoefectyvnosti"* (in Ukrainian) [On establishement of the state institution Energy Efficiency Fund]. http://zakon2.rada.gov.ua/laws/show/1099-2017-%D0%BF. Accessed 23 May 2018.

Cabinet of Ministers of Ukraine (CMU). (2017c). *Pro zatverdzhennia poriadku vykorystannia koshtiv, peredbachenyh u derzhavnomu biudzheti dlia funktsionuvannia Fondu energoefectyvnosti* (in Ukrainian) [On approval of the procedure of the use of funds provided by the state budget for functioning of the Energy Efficiency Fund]. http://zakon2.rada.gov.ua/laws/show/1102-2017-%D0%BF. Accessed 23 May 2018.

Cabinet of Ministers of Ukraine (CMU). (2017d). Pro vykonannia Ugody pro asotsiatsiiu mizh Ukrainoiu, z odniei storony, ta Yevropeiskym Soiuzom, Yevropeiskym spivtovarystvom z atomnoi energii i iihnimy derzhavamy-chlenamy, z inshoi storony (in Ukrainian) [On fulfilment of the association agreement between Ukraine, on the one hand, and the European Union, the European Atomic Energy Community and their member-states, on the other hand]. http://zakon3.rada.gov.ua/laws/show/1106-2017-%D0%BF/paran34#n10. Accessed 23 May 2018.

DTEK. (2017). *DTEK renewables B.V. Ukraine: Trade finance 2017*. http://wwe-conferences.com/wp-content/files/Ms_Nersesova.pdf. Accessed 23 May 2018.

Dukunskyy, I., & Zharikov, A. (2016). *Legal alert: Renewable energy*. http://dlf.ua/en/legal-alert-renewable-energy/. Accessed 23 May 2018.

Eurostat. (2018). *Complete energy balances—Annual data*. http://appsso.eurostat.ec.europa.eu/nui/show.do?dataset=nrg_110a&lang=en. Accessed 23 May 2018.

General Prosecutor's Office (GPO). (2014). *Genprokuraturoiu rozpochato kryminalni provadzhennia za factom zavolodinnia mainom DAT "Chornmornaftogaz"* (in Ukrainian) [General prosecutor's office commenced criminal proceedings after the fact of seizure of assets of state joint stock company Chornmornaftogaz]. https://www.gp.gov.ua/ua/news.html?_m=publications&_c=view&_t=rec&id=145489. Accessed 23 May 2018.

Information and Communication Department of the Secretariat of the Cabinet of Ministers of Ukraine (CMU). (2016b). *Volodymyr Groysman: Over 5 years Ukraine should become a fully energy independent state*. http://old.kmu.gov.ua/kmu/control/uk/publish/article?art_id=249307782&cat_id=244314975. Accessed 23 May 2018.

Institute for 21st Century Energy of U.S. Chamber of Commerce's. (2016). *International Index of Energy Security Risk, 2016 edition*. https://www.globalenergyinstitute.org/sites/default/files/energyrisk_intl_2016.pdf. Accessed 23 May 2018.

Intended Nationally-Determined Contribution (INDC) of Ukraine to a New Global Climate Agreement. (2016). http://www4.unfccc.int/Submissions/INDC/Published%20Documents/Ukraine/1/150930_Ukraine_INDC.pdf. Accessed 23 May 2018.

Interfax. 2018. *Ukraine cuts anthracite consumption by 46% in 2017—Deputy PM Kistion*. https://en.interfax.com.ua/news/general/477295.html. Accessed 23 May 2018.

International Bank for Reconstruction and Development (IBRD)/The World Bank. (2016). *Ukraine country environmental analysis*. https://openknowledge.worldbank.org/bitstream/handle/10986/24971/Ukraine000Coun0ironmental0analysis.pdf?sequence=4&isAllowed=y. Accessed 23 May 2018.

Isachenko, S. (2014, May 19). *Gazogon "Urengoy-Pomary-Uzhgorod" pidryvaly trychi (in Ukrainian)* [Gas pipeline was blown up three times]. Uriadovyi Kurier. http://ukurier.gov.ua/uk/news/gazogon-urengoj-pomari-uzhgorod-pidrivali-trichi/. Accessed 23 May 2018.

Ministry of Economic Development and Trade (MEDTU). (2013). *Nakaz Pro Zatverdzhennia metodychnykh rekomendatsiy schodo rozrakhunku rivnia economichnoi bezpeky Ukrainy* (in Ukrainian) [Order on approval of methodological recommendations on calculation of the level of economic security of Ukraine]. http://www.me.gov.ua/Documents/List?lang=ukUA&tag=MetodichniRekomendatsii. Accessed 23 May 2018.

National Energy and Utilities Regulatory Commission of Ukraine (NEURCU). (2017). *OPIC decided on financing the largest wind farm in the Ukraine at the PPA, prepared NKREKP*. http://www.nerc.gov.ua/?news=7101. Accessed 23 May 2018.

National Institute for Strategic Studies (NISS). (2015). *Donbass and the Crimea: The value of return: Monograph*. Under the general editorship of V. Horbulin, O. Vlasiuk, E. Libanova & O. Lyashenko. Kyiv: NISS. http://en.niss.gov.ua/public/File/englishpublic/Eng_Donbass_Poltor.pdf. Accessed 23 May 2018.

National Police of Ukraine (NPU). (2014). *Arsen Avakov: dyversiia na gazoprovode v Poltavskoy oblasti – chergova sproba z boku Rosii dyskredytuvaty Ukrainu yak partnera v gazoviy sferi* (in Ukrainian) [Arsen Avakov: Sabotage on the pipeline in Poltava oblast—Another attempt of Russia to discredit Ukraine as a partner in the gas sector]. https://www.npu.gov.ua/uk/publish/article/1080372. Accessed 23 May 2018.

National Security and Defense Council of Ukraine (NSDCU). (2016). *Pro udoskonalennia zakhodiv zabezpechennia zakhystu obiektiv krytychnoi inrastruktury* (in Ukrainian) [On improvements of measures on ensuring protection of critical infrastructure objects]. http://zakon2.rada.gov.ua/laws/show/n0014525-16. Accessed 23 May 2018.

Nowak, Z. (2015). *Unused potential: Renewable energy sources in Ukraine*. http://www.pism.pl/files/?id_plik=19601. Accessed 23 May 2018.

Overseas Private Investment Corporation (OPIC). (2017). *OPIC board of directors approves $400 million in financing and political risk insurance for wind power project in Ukraine*. https://www.opic.gov/press-releases/2017/opic-board-directors-approves-400-million-financing-and-political-risk-insurance-wind-power-project-ukraine. Accessed 23 May 2018.

Organisation for Economic Co-operation and Development (OECD). (2012). *Attracting investment in renewable energy in Ukraine*. Private sector development policy handbook. https://www.scribd.com/document/153898969/Attracting-Investment-in-Renewable-Energy-in-Ukraine. Accessed 23 May 2018.

Roberts, A. (2017). *Deterioration of an ecological situation on Donbas*. http://www.youreporter.it/foto_Deterioration_of_an_ecological_situation_on_Donbas?refresh_ce-cp. Accessed 23 May 2018.

Solonyna, Y. (2014). *Minister of ecology: Russian troops are ruining Donbas' environment*. http://euromaidanpress.com/2014/09/11/minister-of-ecology-russian-troops-are-ruining-donbas-environment/#arvlbdata. Accessed 23 May 2018.

State Agency of Energy Efficiency and Energy Saving of Ukraine (SAEE). (2016). *Uriadova programa z energoefectyvnosti dlia pryvatnyh domogospodarstv* (in Ukrainian) [Governmental energy efficiency program for private households]. http://saee.gov.ua/sites/default/files/DG.pdf. Accessed 23 May 2018.

State Agency of Energy Efficiency and Energy Saving of Ukraine (SAEE). (2017a). *Bilche 6 mlrd. hrn. investytsiy zalucheno v ekonomiku Ukrainy zavdiaky programi "teplyh kredytiv"* (in Ukrainian) [More than 6 bln. Hryvnias of investments are attracted to economy of Ukraine as a result of 'warm loans' program]. http://saee.gov.ua/uk/news/2099. Accessed 23 May 2018.

State Agency of Energy Efficiency and Energy Saving of Ukraine (SAEE). (2017b). *Reformy v sferi energoefektyvnsti ta vidnovliuvanoi energetyky: dosiagnennia za 3 roky, initsiatyvy ta ihniy efekt na ekonomiku Ukrainy* (in

Ukrainian [Reforms in the field of energy efficiency and renewable energy: Progress for 3 years, initiatives and their impact on economy of Ukraine]. http://saee.gov.ua/sites/default/files/REPORT_29_08_2017.pdf. Accessed 23 May 2018.

State Agency of Energy Efficiency and Energy Saving of Ukraine (SAEE). (2017c). *Hidroenergetyka* (in Ukrainian) [Hydropower industry]. http://saee.gov.ua/uk/ae/hydroenergy. Accessed 23 May 2018.

State Agency of Energy Efficiency and Energy Saving of Ukraine (SAEE). (2018). *Informatsiia shchodo potuzhnosti ta obsiagiv vyrobnytstva electroenergii obiektamy vidnovliuvanoi elekntroenergetyky, iakym vstanovleno "zelenyi" taryf (stanom na 31.12.2017)* (in Ukrainian) [Information on capacity and volumes of electricity production by renewable energy objects for which "green" tariff is established (as of 31.12.2017)]. http://saee.gov.ua/sites/default/files/IV%202017.pdf. Accessed 23 May 2018.

State Agency on Exclusion Zone Management (SAEZM). (2016). *Chornobyl solar: New opportunities in renewable energy sector.* Presentation. http://dazv.gov.ua/vidnovliuvalna-enerhetyka/vidnovlyuvalna-energetika-u-zoni-vidchuzhennya-novi-mozhlivosti-dlya-sektoru-energetiki.html. Accessed 23 May 2018.

State Reserve Agency (SRA). (2017). *Energetychne Spivtovarystvo shvalylo model MZNN, obranu dlia Ukrainy* (in Ukrainian) [Energy community approved model of minimum stock of crude oil and petroleum products chosen for Ukraine]. https://rezerv.gov.ua/news/energetychne-spivtovarystvo-shvalylo-model-mznn-obranu-dlja-ukrainy. Accessed 23 May 2018.

State Statistics Service of Ukraine (Ukrstat). (2017a). *Naiavnist i stan osnovnykh zasobiv za vydamy economichnoi diialnosti (KVED-2010) za 2016 rik* (in Ukrainian) [Availability and status of fixed assets by types of economic activities (KVED-2010) for 2016)]. http://ukrstat.gov.ua/. Accessed 23 May 2018.

State Statistics Service of Ukraine (Ukrstat). (2017b). *Energetychnyi balans Ukrainy za 2015 Rik (express-edition)* (in Ukrainian) [Energy balance of Ukraine for 2015 (express edition)]. http://ukrstat.gov.ua/. Accessed 23 May 2018.

Sukhodolia, O. (2014). *Zakhyst energetychnoi infrastructury: analiz ukrainskoi zakonodavchoi bazy* (in Ukrainian) [Protection of energy infastructure: Analysis of Ukrainian legal framework]. Analytical note. http://niss.gov.ua/articles/1568. Accessed 23 May 2018.

Ukraine Energy Efficiency Secretariat and Expert Hub, Project of United Nations Development Programme (Energy Efficiency Secretariat). (2017). *Rozvytok vidnovliuvanyh dzerel energii v Ukraini* (in Ukrainian) [Development of renewable energy sources in Ukraine]. http://abc.in.ua/wp-content/uploads/2017/03/Rozvitok-VDE-v-Ukrai-ni.pdf. Accessed 23 May 2018.

United Nations Climate Change (UNFCCC Secretariat). (2012). *Doha amendment to the Kyoto protocol*. https://unfccc.int/files/kyoto_protocol/application/pdf/kp_doha_amendment_english.pdf. Accessed 23 May 2018.

Unigovskiy, L. (2016). *Rynok pryrodnogo gaza poka deliok ot kokurentsii (chast I)* (in Russian) [The natural gas market is far from competitiveness (part I)]. http://hubs.ua/economy/l-unigovskij-ry-nok-prirodnogo-gaza-poka-dalek-ot-konkurentsii-chast-i-66236.html. Accessed 29 December 2017.

Verkhovna Rada of Ukraine (VRU). (2003). *Zakon pro osnovy natsionalnoi bezpeky Ukrainy* (in Ukrainian) [Principles of National Security Act]. http://zakon0.rada.gov.ua/laws/show/964-15. Accessed 23 May 2018.

Verkhovna Rada of Ukraine (VRU). (2010). *Podatkovyi kodeks Ukrainy* (in Ukrainian) [Tax code of Ukraine]. http://zakon2.rada.gov.ua/laws/main/2755-17. Accessed 23 May 2018.

Verkhovna Rada of Ukraine (VRU). (2017a). *Zakon pro Fond Energoefectyvnosti* (in Ukrainian) [Energy Efficiency Fund Act]. http://zakon2.rada.gov.ua/laws/show/2095-19. Accessed 23 May 2018.

Verkhovna Rada of Ukraine (VRU). (2017b). *Zakon pro rynok electrychnoi energii* (in Ukrainian) [Electricity Market Act]. http://zakon3.rada.gov.ua/laws/main/2019-19. Accessed 23 May 2018.

Verkhovna Rada of Ukraine (VRU). (2017c). *Zakon pro otsinku vplyvu na dovkillia* (in Ukrainian) [Environmental Impact Assessment Act]. http://zakon3.rada.gov.ua/laws/main/2059-19. Accessed 23 May 2018.

Verkhovna Rada of Ukraine (VRU). (2017d). *Zakon pro osnovni zasady zabezpechennia kiberbezpeky Ukrainy* (in Ukrainian) [Act on Basic Principles of Ensuring Cybersecurity of Ukraine]. http://zakon0.rada.gov.ua/laws/show/2163-19. Accessed 23 May 2018.

Verkhovna Rada of Ukraine (VRU). (2017e). *Zakon pro vnesennia zmin do deiakyh zakonodavchyh aktiv Ukrainy schodo dereguliatsii v naftogazovoi galuzi* (in Ukrainian) [Act on Amending Certain Legislative Acts of Ukraine Concerning Deregulation in the Oil and Gas Sector]. http://zakon2.rada.gov.ua/laws/show/2314-19. Accessed 23 May 2018.

Verkhovna Rada of Ukraine (VRU). (2018). *Zakon pro strategichnu ecologichnu otsinku* (in Ukrainian) [Strategic Environmental Assessment Act]. http://zakon2.rada.gov.ua/laws/show/2354-19. Accessed 23 May 2018.

World Data Atlas (WDA). (2018). *Ukraine—CO_2 emissions intensity*. https://knoema.com/atlas/Ukraine/CO2-emissions-intensity?origin=knoema.ru. Accessed 23 May 2018.

World Bank, International Comparison Program Database. (2018). *GDP, PPP (constant 2011 international $)*. http://data.worldbank.org/indicator/NY.GDP.MKTP.PP.KD. Accessed 23 May 2018.

Yale Center for Environmental Law & Policy, Yale University, Center for International Earth Science Information Network, Columbia University, etc. (2018). *2018 Environmental Performance Index*. https://epi.envirocenter.yale.edu/downloads/epi2018policymakerssummaryv01.pdf. Accessed 23 May 2018.

Zoï (Denisov, N., Simonett, O., Weir, D., Averin, D.) (Zoï). (2015). *The Ukraine conflict's legacy of environmental damage and pollutants*. https://sustainablesecurity.org/2015/04/21/the-ukraine-conflicts-legacy-of-environmental-damage-and-pollutants-2/. Accessed 23 May 2018.

Zwijnenburg, W. (2017). *Donbas: A ticking toxic time bomb*. https://www.bellingcat.com/news/uk-and-europe/2017/03/30/donbas-ticking-toxic-time-bomb/. Accessed 23 May 2018.

Part II

Solutions for Resource Efficiency: Case Studies

7

Energy Resource Efficiency in the EU: Major Legislative Initiatives

Umut Turksen

Introduction

This chapter outlines the key priority areas in the context of energy efficiency in the European Union (EU) and discusses the relevant legal instruments with a view of critically analysing the performance of the EU Member States (MS) against the benchmarks therein. While it is clear that the EU has been spearheading the energy efficiency and development of renewable energy initiative, there is a lack of unified success across the union.

While energy has always been an important and a volatile field, liberalisation of energy trade in general and energy efficiency in particular at the international level have not always been priorities for states (Cottier et al. 2010). Given the ideological, political and economic divisions, particularly between ex-communist nations that joined the EU and market-oriented EU nations and their respective economic conditions

U. Turksen (✉)
Coventry Law School and Research Centre for Financial and Corporate Integrity, Coventry University, Coventry, UK
e-mail: aa8628@coventry.ac.uk

N. Mouraviev and A. Koulouri (eds.), *Energy Security*,
https://doi.org/10.1007/978-3-030-01033-1_7

and priorities, policy developments were polarised and/or limited not only for energy trade but also for energy efficiency targets and standards.

The concept of 'security of energy supply' (Saga 1995; Barton et al. 2004; Cameron 2007; Alhajji 2008; Turksen 2018) is often used interchangeably with another term: energy security. They have, in the opinion of some, a similar—or even the same—meaning (Maican 2009) but the difference between the two terms or concepts has been discussed in Chapter 1 of this book. While even the courts acknowledge the importance of energy security,[1] there is no comprehensive and holistic definition of energy security in law, which encompasses the interrelation between energy efficiency and energy security. Despite this gap, security of energy supply is currently at the top of the agendas of most EU states (Flynn 2006) and the EU,[2] which makes it an EU security issue. Security of supply is also a cornerstone of European energy policy (Selivestrov 2009) being one of its three main objectives.[3]

Similarly, there is no internationally agreed definition of energy efficiency yet, although energy efficiency is accepted as one of the key elements in bolstering energy security, both in terms of geopolitics and security of supply (Turksen 2018); enhancing economic development and growth; and ensuring a safe, reliable, affordable and sustainable energy system and environment for the future. For example, in Germany and the UK, which are the EU's largest gas markets, improvement of energy efficiency in 2016 resulted in gas savings equivalent to 30% of the EU's total imports from Russia (The International Energy

[1]The Court of Justice of the EU held that energy in the modern economy is "of fundamental importance for a country's existence since not only its economy but above all its institutions, its essential public services and even the survival of its inhabitants depend upon them": *Campus Oil*, Case 72/83—Judgment of the Court of 10 July 1984.

[2]The EU's official view on Energy Security: "*Energy supply security must be geared to ensuring, the proper functioning of the economy, the uninterrupted physical availability at a price which is affordable while respecting environmental concerns. Security of supply does not seek to maximise energy self-sufficiency or to minimise dependence, but aims to reduce the risks linked to such dependence.*": EC Green Paper (2000) and Commission Priority, Energy Union: Making energy more secure, affordable and sustainable, http://ec.europa.eu/priorities/energy-union/index_en.htm.

[3]Among "sustainability" and "competitiveness"—Commission Priority, Energy Union: Making energy more secure, affordable and sustainable, http://ec.europa.eu/priorities/energy-union/index_en.htm.

Agency 2017), which in turn reduced reliance on a dominant supplier (Russian state-owned company Gazprom) and contributed towards energy security. Energy efficiency is also identified as a key factor in increasing GDP (e.g. by savings and generation of jobs) and decreasing the global greenhouse gas emissions. The sectors focusing on enhancing energy efficiency have seen a steady increase in investment globally: in 2016, global investment in energy efficiency increased by 9% to $231 billion, 30% of which (the largest share) came from the EU.

Since its inception, the EU has been characterised as a normative power in international trade relations based on liberal values, rules and norms and the dissemination of these rules beyond its borders (Wagnsson 2010; Youngs 2010; Kuzemko 2014). While the EU has begun to create a fully integrated, regulatory regime for its internal energy market via a number of legal instruments[4] and strives to materialise a multilateral energy trade regime (International Energy Charter 2015), such harmonised and integrated approaches have not yet been fully developed when it comes to energy efficiency and sustainability. This is surprising given the fact that in the 1950s, the predecessor of the EU, the European Community (EC) was founded in the context of energy security and trade, namely coal and nuclear energy. For example, the first legal instrument created for this purpose—establishing the European Coal and Steel Community (ECSC) in 1952[5]—was primarily concerned with the distribution of domestic energy resources among its original six MS. Subsequently, the EC's focus began to shift from internal regulation of coal to external supply of other energy

[4]The EU internal energy market is regulated under the Third Energy Package which consists of two Directives and three Regulations: Directive 2009/72/EC concerning common rules for the internal market in electricity and repealing Directive 2003/54/EC; Directive 2009/73/EC concerning common rules for the internal market in natural gas and repealing Directive 2003/55/EC; and Regulation (EC) No. 714/2009 on conditions for access to the network for cross-border exchanges in electricity and repealing Regulation (EC) No. 1228/2003; Regulation (EC) No. 715/2009 on conditions for access to the natural gas transmission networks and repealing Regulation (EC) No. 1775/2005; Regulation (EC) No. 713/2009 of the European Parliament and of the Council of 13 July 2009 establishing an Agency for the Cooperation of Energy Regulators.

[5]The Treaty itself expired in 2002, but some of its provisions were incorporated into subsequent treaties. For details see http://europa.eu/legislation_summaries/institutional_affairs/treaties/treaties_ecsc_en.htm.

resources (including oil and gas and later, renewables). This shift was not legally reflected in the treaties and the MS were free to determine their relations with non-EC countries (Belyi et al. 2011), which now supply the majority of the EU's energy (particularly oil and gas supplies). For example, the Russian Federation is the dominant energy supplier for the EU (Plebalgs 2009). Russia is reported to have the largest natural gas reserves (about 18%) and seventh largest crude oil reserves in the world; it is also the biggest exporter of oil and gas to the EU, with its supplies accounting for 25% of oil and 33% of gas.[6]

Despite its efforts to increase domestic renewable energy supplies, the EU as a whole remains the world's largest energy importer, importing approximately 55% of its energy supply including nearly 84% of its oil and 64% of its natural gas.[7] The EU's overall dependency on gas imports is expected to rise significantly by 2030[8] with its energy consumption rising by 15% in comparison to the demand in 2000. This is particularly true with respect to the EU's imports of natural gas. Projections show that European gas production is expected to decline (International Energy Outlook 2004) and the reliance on imported natural gas will grow considerably (International Energy Outlook 2005), a fact recognised already in EU Council Directive 2004/67/EC.[9]

The lack of legal mandate on the EU's scrutiny of external energy supplies and energy efficiency measures was only partially justifiable because, although energy reserves were sufficient at that time, the future dependence on external energy resources was foreseeable. Importantly, reliance on importing resources to the EU has had significant consequences for its energy efficiency standards, targets and priorities set out

[6]See, www.energy.eu/#dependence and Eurogas statistics at www.eurogas.org. Russia provides 100% of gas imports of at least seven Member States of the EU. The EU Commission, EU Energy Policy Data, SEC (2007), p. 12.

[7]European Commission, Market Observatory for Energy, *Key Figures*, June 2011. See, http://ec.europa.eu/energy/observatory/eu_27_info/doc/key_figures.pdf. In 2011, oil made up about 37%, coal nearly 18%, and nuclear energy 12% of the EU primary energy supply.

[8]European Commission (2000). Towards a European Strategy for the Security of Energy Supply. (Green Paper) COM(2000) 769 Final.

[9]Council Directive 2004/67/EC of 26 April 2004 concerning measures to safeguard security of natural gas supply, OJ L 127, 29/04/2004, Preamble, para. 13.

by the Road Map (Roadmap 2011) because EU standards and policy cannot be applied extra-territorially, particularly in relation to exploration, extraction, processing and transport. Accordingly, this chapter is informed by the EU's binding legal instruments (*acquis communautaire*) that underpin the EU's *Single Energy Market* and govern the trading, transport and sale of energy products in the EU Member States.[10]

Despite the inherent lack of specific international legal mandate in the context of energy resource efficiency and given the fact that over 68% of the world's energy use is not covered by efficiency codes or standards,[11] the EU has achieved considerable progress in establishing a common ground for ambitious aims and objectives and continues to be the major driving force behind global efforts in this area.[12] The current EU Commission states that efficient and fully integrated energy networks are the backbone of the single market[13] and 'energy security dimension is recognised as one of the cornerstones of the Energy Union strategy, a key political priority of the Juncker Commission'.[14]

The EU consumes 12% of the world's energy output annually and the EU's energy system is still underperforming as a whole because of the different success rates in energy efficiency. The EU Commission reported that the EU 'has the world's highest net imports of resources per person, and its open economy relies heavily on imported raw materials and energy'.[15] The vision of the EU for energy efficiency is

---

[10]The Third Energy Package, supra note 4. Also see, the EU Commission, A fully-integrated internal energy market, https://ec.europa.eu/commission/priorities/energy-union-and-climate/fully-integrated-internal-energy-market_en.

[11]The International Energy Agency, Energy Efficiency 2017, https://www.iea.org/publications/freepublications/publication/Energy_Efficiency_2017.pdf.

[12]For example, ISO 50001—a global standard for energy management developed by the International Organization for Standardization in 2011—grew to nearly 12,000 in 2015, 85% of which were in in the EU. Ibid.

[13]EU Commission, Single Market Act II—Together for New Growth, COM(2012) 573 final. 03.10.2012.

[14]EU Commission—Press Release, Towards Energy Union: Sustainable energy security package, Brussels, 16 February 2016, http://europa.eu/rapid/press-release_IP-16-307_en.htm.

[15]Communication from the Commission to the European Parliament, the Council, the European Economic and Social Committee and the Committee of the Regions Roadmap to a Resource Efficient Europe, COM/2011/0571 Final, http://eur-lex.europa.eu/legal-content/EN/TXT/?uri=CELEX:52011DC0571.

contained and articulated in the EU Roadmap to a Resource Efficient Europe (Roadmap 2011), which outlines the policy targets and actions to produce more value with less input, use resources (including energy) in a sustainable way and manage them more efficiently throughout their life cycle.[16] The fact that this vision was initially articulated as a 'communication' from the EU Commission (which is not legally binding on the EU Member States)[17] indicates that while there is a political will at the EU level, there are different capabilities and fragmented approaches to energy resource efficiency across the EU at the MS level. For example, as illustrated in Fig. 7.1, the production and use of renewable energy across Europe vary significantly.

While all MS of the EU belong to the developed countries category,[18] the EU's energy efficiency policy cannot undermine the members' respective and unique developmental needs. It tries to strike a balance between environmental and sustainability priorities, on the one hand, and achieve impact on the member states' developmental needs, on the other.[19] Since the 2011 Roadmap, there have been numerous initiatives, including the EU's Circular Economy Package 2015 and the subsequent Action Plan 2017,[20] which were designed to provide substantive methods to achieve resource efficiency.

In part one, this chapter provides an overview of the general policy (the Roadmap), the legal framework and envisaged targets for EU

---

[16]Ibid. The EU Parliament has supported the Roadmap by issuing a resolution, namely the European Parliament resolution of 24 May 2012 on a resource-efficient Europe (2011/2068(INI)).

[17]For the list of sources of EU Law see, http://www.europarl.europa.eu/atyourservice/en/displayFtu.html?ftuId=FTU_1.2.1.html.

[18]The UN Country Classifications, http://www.un.org/en/development/desa/policy/wesp/wesp_current/2014wesp_country_classification.pdf.

[19]A typical example in this regard is Directive 2009/28/EC of the European Parliament and of the Council of 23 April 2009 on the promotion of the use of energy from renewable sources which states that "Cyprus and Malta, due to their insular and peripheral character, rely on aviation as a mode of transport, which is essential for their citizens and their economy. As a result, Cyprus and Malta have a gross final consumption of energy in national air transport which is disproportionately high, i.e. more than three times the Community average in 2005, and are thus dispropor-tionately affected by the current technological and regulatory constraints".

[20]Communication from the Commission to the European Parliament, the Council, the European Economic and Social Committee and the Committee of the Regions Closing the Loop—An EU action plan for the Circular Economy, COM/2015/0614 Final, http://eur-lex.europa.eu/legal-content/EN/TXT/?uri=CELEX:52015DC0614.

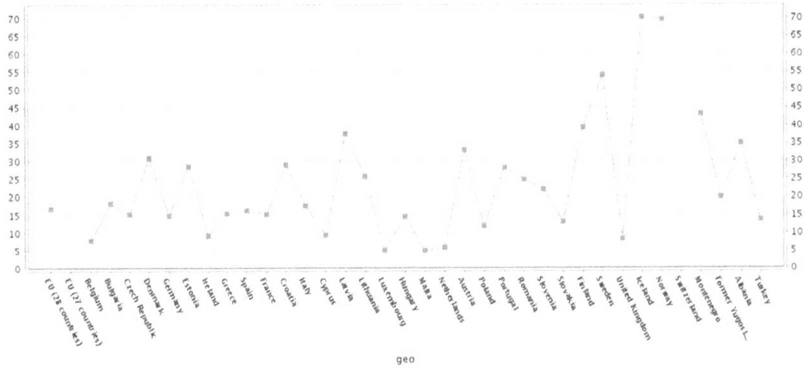

Fig. 7.1 Share of renewable energy in gross final energy consumption across Europe (*Source* Eurostat 2018)

Member States in the context of energy resource efficiency. This focus derives from the fact that, although originally it was not a legally binding regime (as part of the EU's *acquis communautaire*), the energy efficiency Roadmap and the subsequent policies inspired by this programme have been the driving force behind some of the main legal instruments (e.g. Energy Efficiency Directive 2012),[21] significant improvements and positive results pertaining to energy efficiency. These developments in turn give the EU Member States the impetus to comply with legally binding obligations, which are discussed below. Part two of the chapter analyses the key legal instruments and achievements, or the lack thereof, in relation to the energy efficiency targets set out by the EU.

The Roadmap to a Resource Efficient Europe

The EU's envisaged Energy Union framework is divided into five dimensions that comprise both legal and policy measures:

---

[21]Directive 2012/27/EU of the European Parliament and of the Council of 25 October 2012 on energy efficiency, amending Directives 2009/125/EC and 2010/30/EU and repealing Directives 2004/8/EC and 2006/32/EC, http://eur-lex.europa.eu/LexUriServ/LexUriServ.do?uri=OJ:L:2012:315:0001:0056:en:PDF.

- Energy security;
- The internal single energy market;
- Energy efficiency;
- Decarbonisation; and
- Research, innovation and competitiveness.

The Roadmap to a Resource Efficient Europe (COM(2011) 571), which encompasses all of these elements, has put forward a long-term strategy and a holistic approach to resource efficiency by proposing EU-wide methods and targets to increase productivity and economic growth without compromising the environment. The framework provided by the Roadmap also sets out a vision for the structural and technological changes needed in order to meet the targets by 2050, with a number of milestones to be reached by 2020 (also known as Energy 2020 Goals).[22] The key thematic areas of the Roadmap can be summarised under the following titles:

- Challenges and opportunities for Europe;
- Making Europe resource efficient;
- Transforming the economy;
- Supporting research and innovation; and
- Priority sectors.

These themes are discussed below.

Challenges and Opportunities for Europe

The Roadmap can be viewed as part of the global effort to achieve a transition towards a green economy, which recognises sustainability as the foundation of its efficiency strategy. This requires a fundamental

[22]For example, it is envisaged that by 2020, there will be at least 20% reduction in greenhouse gas emissions compared to 1990 (30% if international conditions are right, European Council, 10–11 December 2009); saving of 20% of EU energy consumption compared to projections for 2020; and 20% share of renewable energies in EU energy consumption, 10% share in transport.

transformation in energy, industry, agriculture, fisheries and transport systems and in producer and consumer behaviour. The transformative framework and the means to materialise it in the medium- and long-term aim to create 'a playing field, where innovation and resource efficiency are rewarded, creating economic opportunities and improved security of supply through product redesign, sustainable management of environmental resources, greater reuse, recycling and substitution of materials and resource savings' (Roadmap 2011, para. 1). The Roadmap also points to some stark statistics and potential scenarios:

'In the EU, each person consumes 16 tonnes of materials annually, of which 6 tonnes are wasted, with half going to landfill' (ibid.).

It predicts that 'if we carry on using resources at the current rate, by 2050 we will need, on aggregate, the equivalent of more than two planets to sustain us, and the aspirations of many for a better quality of life will not be achieved' (ibid.). The EU underlines the fact that there is a direct correlation between economic prosperity and sustainability of natural capital and biodiversity, which underpins our ecosystems including those from which energy is produced. It is envisaged that by '2020 the loss of biodiversity in the EU and the degradation of ecosystem services will be halted and, as far as feasible, biodiversity will be restored' (Roadmap 2011, para. 4.2). In this context, the Roadmap prioritises agriculture and fisheries and gives scant attention to energy. However, when addressing the use of water, the Roadmap recognises the importance of water for human health, agriculture, tourism, industry, transport and energy, and recognises the fact that a reduction in 'water availability has a critical impact on hydropower and cooling of nuclear and thermal power stations' (Roadmap 2011, para. 4.4).

The EU reports that about 20–40% of Europe's water is wasted and water efficiency could be significantly increased through technological improvements. The Roadmap asserts that sustainable management of water resources requires close coordination with agriculture, transport, regional development and energy policies as well as effective and fair water pricing as required by the Water Framework Directive.[23]

[23]Directive 2000/60/EC of the European Parliament and of the Council of 23 October 2000 establishing a framework for Community action in the field of water policy; http://eur-lex.europa.eu/legal-content/EN/TXT/?uri=CELEX:32000L0060.

Accordingly, the Roadmap sets a number of targets which mainly focus on quality, quantity and use of water but at first glance, does not set any explicit targets in relation to the use of water for energy extraction and production, such as hydroelectric and hydro-fracturing (fracking). It could be argued that this omission is remedied when the Roadmap addresses the use of land and mentions that when land is used, there is often a 'trade-off between various social, economic and environmental needs (e.g. housing, transport infrastructure, energy production, agriculture, nature protection)' (Roadmap 2011, para. 4.6). There is a clear emphasis on the requirement for Strategic Environmental Assessment (SEA, also known as environmental impact assessment)[24] and the need to 'address the indirect land use change resulting notably from the renewable energy policy' (ibid.). The Roadmap provides specific targets for member states, including to 'better integrate direct and indirect land use and its environmental impacts in their decision making and limit land take and soil sealing to the extent possible' (ibid.). This is an area in which the EU Commission has been proactive in utilising its enforcement powers. For example, in the case of the construction of the South Stream Pipeline project, the Bulgarian government failed to conduct a SEA and inform the EU Commission. Consequently, the Commission took enforcement action under Article 258 of the Treaty on the Functioning of the EU (TFEU),[25] which stopped the multibillion euro project and eventually led to the unilateral termination of the project by President Putin (Turksen 2018, 32–33). Although such actions are rare, the EU is prepared to take action to bring about compliance when essential interests of the EU are at stake and when the legal provisions allow.

Importantly, legal enforcement is only one of the means of bringing about change and compliance, and not the ideal way. A cultural change

[24]Directive 2011/92/EU of the European Parliament and of the Council of 13 December 2011 on the assessment of the effects of certain public and private projects on the environment Text; http://eur-lex.europa.eu/legal-content/EN/TXT/?uri=celex%3A32011L0092.

[25]European Commission, 'Internal energy market: Commission refers Bulgaria, Estonia and the United Kingdom to Court for failing to fully transpose EU rules' (Press Release Database, 24 January 2013); http://europa.eu/rapid/press-release_IP-13-42_en.htm.

needs to start with a bottom-up approach that is supported by incentives and holistic regulation. The EU recognises that there are significant differences in the performance of member states in implementing energy efficiency measures,[26] especially in the context of nature conservation, waste and water management. The EU estimates the costs of failing to implement current legislation to be around 50 billion EUR per year,[27] yet it is not clear what it would cost to bring about coherent implementation across the EU.

Making Europe Resource Efficient

The Roadmap strives to achieve economic development with less energy consumption via new technologies whereby all environmental assets (within and outside the EU) that benefit the EU are secure and managed within their maximum sustainable yields. In order to achieve this, the Roadmap puts forward two main indicators to measure progress:

- A provisional lead indicator—*resource productivity*—to measure the principal objective of this Roadmap: that of improving economic performance while reducing pressure on natural resources.
- A series of complementary indicators on key natural resources, such as water, land, materials and carbon, that will take account of the EU's global consumption of these resources.

In addition, the Roadmap recognises the negative impact of technical, legal and social barriers on energy efficiency and therefore proposes various incentives for changes in production and consumption, which require, inter alia:

[26]The EU has set up a The Resource Efficiency Scoreboard which presents indicators covering themes and subthemes of the Roadmap to a Resource Efficient Europe and aims to monitor implementation across the EU: http://ec.europa.eu/eurostat/web/europe-2020-indicators/resource-efficient-europe.

[27]EU Commission, Financing Energy efficiency, https://ec.europa.eu/energy/en/topics/energy-efficiency/financing-energy-efficiency.

- Investigating the functioning of markets and possibly revising prices, taxes and subsidies that do not reflect the real costs of resource use and lock the economy onto an unsustainable path.
- Encouraging more long-term innovative thinking in business, finance and politics that leads to the uptake of new sustainable practices, stimulates breakthroughs in innovation and develops forward-thinking and cost-effective regulation.
- Carrying out the research to fill the gaps in knowledge and skills and providing the right information and training.
- Dealing with concerns regarding international competitiveness and seeking to achieve consensus with international partners to move in a similar direction.

Transforming the Economy

The EU predicts that the process and consequences of the transformation of the economy to one that is energy-efficient will increase competitiveness and act as a driver for growth and job creation through innovation and commercialisation of new technology. The EU persistently emphasises sustainable production and consumption of resources, which includes energy. This is an area where the EU makes reference to existing legal provisions with direct effect, whereby the EU's Single Market and its instruments have an important role in setting the framework for markets to reward greener products through voluntary and mandatory measures, such as the EU's Lead Market Initiatives and the Ecodesign Directive.[28] Importantly, the Roadmap recognises the need to change the mindset of citizens regarding their consumption habits and the importance of transitioning to a low-carbon economy. For example, cost savings made from improving the efficiency of technology can actually induce people to consume more; this is known as the rebound effect. Accordingly, there is a need to anticipate and account

[28]Directive 2009/125/EC of the European Parliament and of the Council of 21 October 2009 establishing a framework for the setting of eco-design requirements for energy-related products, http://eur-lex.europa.eu/legal-content/EN/TXT/?uri=CELEX:32009L0125.

for such behaviours when developing policy and setting targets. In this regard, the key milestone set by the Roadmap is that by 2020 citizens and public authorities should have the right incentives to choose the most resource-efficient products and services and business investments in efficiency are rewarded.

In tandem with this priority, the EU has set out its Green Public Procurement (GPP) standards for products with significant environmental impacts (environmental footprint and eco-design). The EU also assessed where GPP could be linked to EU-funded projects, as well as promoted joint procurement and networks of public procurement officers in support of GPP. The GPP model has been used in various ventures including the use of food, paper, clothing, cleaning products and lighting.[29] In the procurement of such ventures, green considerations, benchmarked against international and EU best practices, are required in the tender. For example, the Clean Streets in Barcelona Project was put out to tender with numerous energy efficiency and sustainability requirements:

- The contract was divided into four parts in order to facilitate participation of small and medium-sized enterprises in the tender.
- Service vehicles had to comply with EURO 5 and operate on non-contaminating combustible fuels and/or renewables, such as biofuels (bioethanol B-85, biodiesel or biogas) or use electric vehicles or hybrid cars.
- Contractors were required to implement actions for reducing energy consumption in the course of the contract, for example for lighting and heating, and to use clean energy, for example from solar panels.

Similarly, in the Green Electricity and Vehicles Project involving 120 public authorities in Slovenia, a number of efficiency specifications were built in the tender. For example, at least 30% of the electricity supplied had to be produced from renewable sources and all vehicles had to meet the EURO 5 emissions standard or equivalent.

---

[29]EU Commission, GPP—A Collection of good practices, http://ec.europa.eu/environment/gpp/pdf/GPP_Good_Practices_Brochure.pdf.

From an entrepreneurship and business point of view, these priorities represent a great opportunity for innovation and transformation of products and services. The EU has set the target for 2020, when environmentally harmful subsidies must be phased out. In this context, the EU also recognises the importance of green tax reforms that consist of increasing the share of environmental taxes, while reducing others in order to ensure more environmentally-friendly consumption.

Supporting Research and Innovation

In this area the EU has been undertaking a holistic and incentivised approach in order to better understand energy efficiency, its sources and constraints and reduction in the use of resources. The EU acknowledges the requirement for a comprehensive and credible knowledge base about how natural systems react to different pressures exerted on them. The priority thematic areas are as follows: meeting resource efficiency objectives; supporting innovative solutions for sustainable energy, transport and construction; management of natural resources; preservation of ecosystem services and biodiversity; resource efficient agriculture and the wider bio-economy; environmentally-friendly material extraction; recycling, reuse, substitution of environmental impact of certain materials; smarter design, green chemistry and lower impact, biodegradable plastics.

Priority Sectors

With reference to empirical evidence and data, the Roadmap lists three key sectors—food, buildings and transport—that rely on energy and are responsible for the majority of negative environmental impacts and therefore must be managed by long-term strategies. For example, food and drink chains in the EU cause 17% of greenhouse emissions and 28% of material resource use. In relation to these key sectors, the term 'renewables' is only found with reference to construction and use of buildings, although the EU aims to strengthen its existing policies for energy efficiency and utilisation of renewable energy. The EU makes a particular reference to SMEs that make up the majority of the construction companies in the EU and that need urgent investment and training to acquire resource-efficient construction methods and practices. The amount of investment needed for meeting the renewable energy targets in the EU is estimated at 1 trillion EUR from 2015 to 2030 (Bloomberg 2014). While the EU has made considerable investment in renewables and leads the global effort in renewable energy investment per capita, its share in total renewables investment has been declining, from almost one half in 2010 to less than one-fifth in 2015,[30] which is the consequence of stronger competition in investment in renewables globally.[31]

The Roadmap sets out an ambitious milestone, that by 2020, the renovation and construction of buildings and infrastructure will be made to high resource efficiency levels, whereby 'all new buildings will be nearly zero-energy and highly material efficient, and policies for renovating the existing building stock will be in place so that it is cost-efficiently refurbished at a rate of 2% per year and 70% of non-hazardous construction and demolition waste will be recycled' (Roadmap 2011, para. 3(a)). In regard to mobility and transport, a similar string of policy targets is put forward with a less ambitious target of a 1% average reduction of greenhouse emissions per annum.

Having outlined the main focus and the milestones in the Roadmap, it is important to consider the legal framework that underpins energy efficiency in the EU. While there are numerous sector-specific energy efficiency rules and regulations, the key EU provisions, designed to harmonise and standardise energy efficiency and increase renewables, can be found in two separate directives as discussed in the next section.[32]

---

[30]Proposal for a Directive of the European Parliament and of the Council on the promotion of the use of energy from renewable sources (recast), COM/2016/0767 final/2—2016/0382 (COD), Section 1.1: http://eur-lex.europa.eu/legal-content/EN/TXT/?uri=CELEX:52016PC0767R%2801%29#footnoteref2.

[31]Ibid.

[32]Note that there are sector specific legal instruments created to improve energy efficiency, including the Energy Efficiency and Energy Performance of Buildings Directives, the EU Emission Trading System proposal of July 2015 and the proposed Effort Sharing Regulation, the Land Use, Land Use Change and Forestry Regulation (LULUCF) of July 2016. However, it is not possible to analyse these within the scope of this chapter.

The Energy Efficiency Directive (EED) and Renewable Energy Directive (RED)[33]

Part two of this chapter focuses on the main legal drivers for energy efficiency in the EU. It is worth noting that the EU has the necessary competence and tools to set rules that promote energy efficiency and increased utilisation of renewable energy.[34] According to Article 288 of the TFEU[35] there are five types of secondary EU legislation, namely:

- regulations
- directives
- decisions
- recommendations
- opinions[36]

While directives are classified as secondary sources in EU Law (and are binding),[37] they are addressed to the MS and created with a focus on the desired result. The directives often provide a degree of discretion (and sometimes certain exemptions) to the MS as to how they would achieve these. Accordingly, MS incorporate directives via national legal instruments, subject to their respective legislative frameworks and rules. Typically, directives are created for policy areas in which total consensus and a unified approach could not be agreed at the EU Council of

[33]Directive 2009/28/EC of the European Parliament and of the Council of 23 April 2009 on the Promotion of the Use of Energy from Renewable Sources and Amending and Subsequently Repealing Directives 2001/77/EC and 2003/30/EC, http://Eur-Lex.Europa.Eu/Legal-Content/EN/ALL/?Uri=Celex%3A32009L0028.

[34]Policy areas including sustainable development, protection of the environment and improvement of citizens' health, creation of jobs, economic growth, reinforcement of energy security all fall within the remit of the EU and Article 194 of the Treaty on the Functioning of the European Union, has conferred the competence to legislate these subject areas including energy.

[35]http://eur-lex.europa.eu/legal-content/EN/TXT/?uri=celex%3A12012E288.

[36]Note that Recommendations and opinions shall have no binding force.

[37]Just as with a domestic legal system, there are a variety of sources of EU law and these create what is called a 'hierarchy of norms'. Some are more important than others, some give the authority by which those others are created.

Ministers. This is the context of energy efficiency in the EU, whereby there are common targets set by directives and the MS can choose methods to achieve them. The rationale behind the EED and RED is also reflected in the title of the EU Commission's research organisation: Energy Efficiency and Renewables Unit.[38]

The Energy Efficiency Directive

According to Article 2(4) of the 2012 EED, energy efficiency means 'the ratio of output of performance, service, goods or energy, to input of energy' and Article 2(6) defines energy efficiency improvement as an increase in energy efficiency as a result of technological, behavioural and/or economic changes. The EED requires the member states to implement a number of measures that are referred to as policy measures. They encompass regulatory, financial, fiscal, voluntary or information-provision instruments, formally established and given effect in a MS to create a supportive framework, requirement or incentive for market actors to provide and purchase energy services and/or undertake other energy efficiency improvement measures.[39]

The key aims and targets of the EED can be summarised as follows:

- Energy distributors or retail energy sales companies have to achieve 1.5% energy savings per year through the implementation of energy efficiency measures.
- EU countries can opt to achieve the same level of savings through other means, such as improving the efficiency of heating systems, installing double-glazed windows or insulating roofs; this is a good example of the discretion and flexibility conferred to the MS under the Directive.
- The public sector in EU countries should purchase energy efficient buildings, products and services.

---

[38]Such strategic approach is also evident in the United States of America, where the relevant organ is titled, the US Department of Energy's Office of Energy Efficiency and Renewable Energy.

[39]Article 2(18) of the EED (2012).

- Governments in EU countries must carry out annual energy efficient renovations on at least 3% (by floor area) of the buildings they own and occupy.
- Energy consumers should be empowered to better manage consumption. This includes easy and free access to data on consumption through individual metering.
- National incentives for SMEs should undergo energy audits.
- Large companies will make audits of their energy consumption to help them identify ways to reduce it, e.g. mandatory energy efficiency certification should accompany the sale and rental of buildings.
- Monitoring of efficiency levels in new energy generation capacities should take place.
- Minimum energy efficiency standards and labelling for a variety of products, such as eco-design for boilers, household appliances, lighting and televisions, should be set.
- The preparation of National Energy Efficiency Action Plans should be done every three years by EU countries.
- Protecting the rights of consumers to receive easy and free access to data on real time and historical energy consumption should be ensured.
- The rollout of 200 million smart meters[40] for electricity and 45 million for gas should take place by 2020.[41]

The EU Commission not only monitors and reports on energy efficiency progress across the EU[42] but also can invoke Article 258 TFEU for enforcement action (also known as infringements procedure) against

---

[40]A smart metre can digitally send metre readings to the energy supplier for more accurate energy bills and provide detailed information to consumers so that they can better understand energy usage and potential savings to be made.

[41]EU Commission, Energy Efficiency, https://ec.europa.eu/energy/en/topics/energy-efficiency. Articles 17 and 23 of Directive 2009/28/EC require the Commission to report biennially to the European Parliament and the Council on the progress achieved in Renewable Energy development in the EU and Member States, and on the EU biofuel sustainability.

[42]The latest progress report can be found here: Report from the Commission to the European Parliament and the Council 2017 assessment of the progress made by Member States towards the national energy efficiency targets for 2020 and towards the implementation of the Energy Efficiency Directive as required by Article 24(3) of the Energy Efficiency Directive 2012/27/EU, http://eur-lex.europa.eu/legal-content/EN/TXT/?qid=1511978095545&uri=COM:2017:687:FIN.

MS for failing to fulfil their obligations.[43] However, to date, there has not been any enforcement action in the context of EED targets and obligations.

The European Commission argues that heating and cooling in buildings and industry account for half of the EU's energy consumption, 84% of which is generated by fossil fuels, while only 16% is generated from renewable energy. Assessment of the progress made by MS towards the national energy efficiency targets for 2020 and towards the implementation of the EED gives a detailed picture of the present state of affairs.[44] In the past few years MS have increased their share of renewables, which contributed to a reduction in primary energy consumption as most sources of renewable energy (excluding biomass and municipal waste) are defined as having 100% transformation efficiency. However, this increase is not close to meeting the current energy consumption needs of the EU. Currently, only 20% of the electricity in the EU is generated from renewable resources.[45] It is envisaged that by 2050, about 55% of gross energy consumption will come from renewables.[46] In order to achieve this, the EU needs significant investment and development not only in terms of renewable technology but also in terms of enhanced integration and connectivity in its energy network and storage capacity. For example, the production and surplus of renewable energy in one part of the EU without sufficient interconnection capacity diminishes the usage and efficiency of such resource elsewhere. While there are large-scale solar power projects across the EU, e.g. in Greece, without interconnections between production, transfer and end-user

[43]http://ec.europa.eu/environment/legal/law/procedure.htm.

[44]Report from the Commission to the European Parliament and The Council, Assessment of the progress made by Member States towards the national energy efficiency targets for 2020 and towards the implementation of the Energy Efficiency Directive 2012/27/EU as required by Article 24 (3) of Energy Efficiency Directive 2012/27/EU, COM(2015) 574 final, https://ec.europa.eu/energy/sites/ener/files/documents/2a_EE%20progress%20report%20-%20CSWD%20part%201.pdf.

[45]Communication from the Commission to the European Parliament, the Council, the European Economic and Social Committee and the Committee of the Regions Energy Roadmap 2050, http://eur-lex.europa.eu/legal-content/EN/ALL/?uri=CELEX%3A52011DC0885.

[46]Ibid.

facilities these development projects will have limited value. The Road Map states that by 2020 an overall increase of interconnection capacity of 40% will be required.[47] Given that the EU has not managed to fully eliminate 'energy islands' against a deadline of 2015, as envisaged, it remains unclear whether the EU will be able to meet its interconnectivity targets. It should be noted that in 2013, the fragmented nature of the EU energy market, as a result of 'insufficient interconnections between national energy networks and…the suboptimal utilisation of existing energy infrastructure', was acknowledged by the Energy Infrastructure Regulation.[48]

Another challenge is to drive the cost of renewables down and offer consumers renewable energy at a competitive price. Even if this is achieved, it does not guarantee that a consumer will have a choice. For example, despite the fact that offshore wind power costs around 57.00 GBP/MWh in the UK, a more expensive alternative is being imposed on consumers via the construction of Hinkley Point C nuclear power station, at a cost of 92.50 GBP/MWh (Grimwood 2018).[49] Thus, there are non-economic forces at play, which determine what kind of energy source consumers use and what impact on production efficiency this might have.

Furthermore, existing renewable technologies require improvement in their efficiency, availability and volume across the EU, e.g. more efficient solar panels, larger wind turbines (Pomerantz 2017).[50] This requires ongoing investment in technological advancement until such technologies mature, become profitable and are available at a local level. Once the EU is able to have a smart energy distribution hub and networks (Alikhanzadeh and Taylor 2014),[51] only then can

[47]The Road Map, para. 3.2 (b), supra note 30.

[48]See, page 1, para. 8 of the Regulation (EU) No. 347/2013 on guidelines for trans-European energy infrastructure and repealing Decision No. 1364/2006/EC and amending Regulations (EC) No. 713/2009, (EC) No. 714/2009 and (EC) No. 715/2009.

[49]If built on time by 2025, Hinkley Point C is projected to cost £19.6 billion.

[50]For example, bigger wind turbines have proven to be more cost effective and efficient.

[51]For example, connecting the North energy resources (mainly gas) and the renewable energy resources form the South (mainly gas and wind power), requires a holistic and more integrated view on transmission, distribution and storage capacities. A Pan-European Electricity Highways System is proposed to be completed by 2050, yet there has been limited progress in this regard.

locally produced renewable energy be distributed and used efficiently. Furthermore, increasing energy supply from indigenous sources (e.g. wind and solar) is likely to ensure security, in a geopolitical sense, by reducing reliance on Russian energy.

While the EED has been instrumental in steering MS to address energy efficiency by putting in place more holistic policies and regulations, previous EU policies that addressed energy efficiency in specific sectors should be acknowledged. For example, following the EU's Energy Performance of Buildings Directives in 2002 and 2010,[52] the Netherlands and Germany increased the strength of residential building codes by more than 60% between 2000 and 2016 and the UK by more than 55% (Brilhante and Skinner 2014). These achievements were a result of transposition of the EU directives into MS by mandatory regulations and energy efficiency policies, such as building energy codes; minimum energy performance standards for lighting, appliances and buildings; fuel economy standards for vehicles; and sectoral standards, e.g. mandatory energy intensity targets for an industry. Such mandatory codes and regulations also incorporated obligations for energy utility providers, requiring them to deliver energy efficiency outcomes.[53]

While a good degree of harmony among the EU MS has been achieved in terms of the implementation of various energy efficiency-related directives, there are still differences in the coverage of these energy efficiency codes and standards across the EU (Rosenow et al. 2016).

Energy Efficiency in the EU

There are four main types of energy efficiency policies in the EU (Ricardo 2014):

---

[52]Directive 2010/31/EU of the European Parliament and of the Council of 19 May 2010 on the energy performance of buildings, http://eur-lex.europa.eu/legal-content/EN/ALL/?uri=celex:32010L0031.

[53]The EU Commission, Clean Energy for all Europeans—Good practice in energy efficiency (2016), https://ec.europa.eu/energy/sites/ener/files/documents/good_practice_in_ee_-web.pdf.

- Energy Efficiency Obligation Schemes (EEOS)[54];
- Financial schemes or fiscal incentives;
- Energy or CO_2 taxes; and
- Regulations or voluntary agreements.

However, the differences in policy coverage and performance of energy efficiency among MS continue to impact the EU's ranking in the IEA's Efficiency Policy Progress Index. For example, Bulgaria's persistent under-achievement brings down the EU's overall ranking.[55]

There are significant differences even between the well-established MS regimes in optimising their energy consumption, as Fig. 7.2 shows. Despite the mixed degree of success among the MS, in the short-term, energy efficiency improvements have strengthened energy security in the EU by reducing daily gas consumption. In this context, a more specific legal instrument was created, namely the 2010 EU Security of Gas Supply Regulation.[56] This regulation put in place a common security indicator, the N-1 standard, which refers to a country's ability to maintain supply to end-users even when a critical piece of gas supply infrastructure is disrupted, such as a pipeline or storage unit. The N-1 standard was created to ensure that, in the case of a disruption of energy supply, the country's remaining available gas infrastructure would be able to meet daily demand if and when there is exceptionally high demand. Those EU member states that use the most energy successfully implemented the N-1 standard. For example, the N-1 indicator

---

[54]EEOS are identified to be the most important type of policy in terms of energy savings whereby 34% of the expected cumulative energy savings across all Member States are to be generated from the implementation of EEOS. Ibid.

[55]The International Energy Agency, Energy Efficiency (2017), p. 48. https://www.iea.org/publications/freepublications/publication/Energy_Efficiency_2017.pdf.

[56]Regulation (EU) No. 994/2010 of the European Parliament and of the Council of 20 October 2010 concerning measures to safeguard security of gas supply and repealing Council Directive 2004/67/EC. Note that Regulation No. 994/2010 was repealed by Regulation (EU) 2017/1938 of the European Parliament and of the Council of 25 October 2017 concerning measures to safeguard the security of gas supply: https://eur-lex.europa.eu/legal-content/en/TXT/?uri=CELEX:32017R1938.

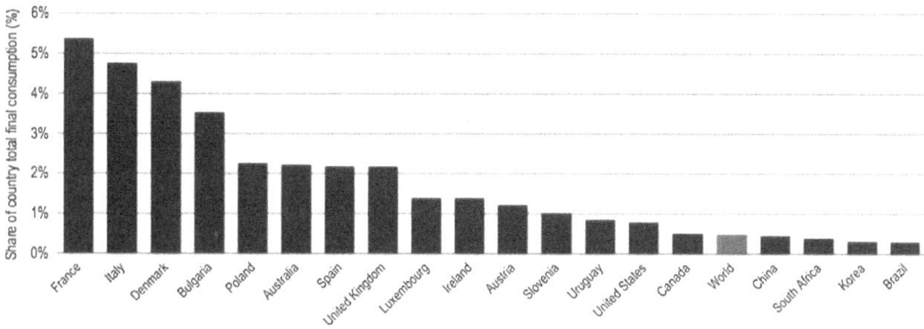

Fig. 7.2 Energy savings in 2016 since 2005, as a percentage of national final energy consumption (*Source* The International Energy Agency, Energy Efficiency 2017, p. 49)

in Germany is 180%, whereas it is 130% in France and 110% in the UK.[57] These achievements are mostly due to gains in efficiency. If there had been no improvement in energy efficiency since 2000 in Europe's three largest gas markets, peak daily gas demands in 2012 would have been higher, reducing the N-1 indicators.[58]

Normative and binding legal tools in the energy efficiency field have been instrumental in creating a common vision and strategy for the EU. However, owing to different geographical, economic, infrastructural and developmental needs, the MS show varying degree of success in increasing their energy efficiency. In addition, while the legal consensus has been put in place, integration of key parts—changing energy mix, transfer of technology and linking of energy networks across the EU—is yet to be achieved.

[57]The European Network of Transmission System Operators for Gas, Security of Gas Supply, 2017; https://www.entsog.eu/publications/security-of-gas-supply.

[58]Ibid.

The Renewable Energy Directive[59]

Renewable energy is recognised as a significant contributor to energy efficiency and security. Renewable energy technologies and sources include hydropower, wind power, solar power, marine energy, geothermal energy, heat pumps, biomass and biofuels. The RED was designed to achieve several goals: an increase in the use of renewable energy so that the EU can meet its obligations under international law, e.g. those agreed under the Kyoto Protocol to the United Nations Framework Convention on Climate Change 2012 and beyond[60]; promotion of energy security; promotion of technological development and innovation; and providing opportunities for employment and regional development. One of the most important aspects of the RED is the *renewable energy obligation*,[61] which means putting in place a national support scheme requiring:

a. energy producers to include a given proportion of energy from renewable sources in their production.
b. energy suppliers to include a given proportion of energy from renewable sources in their supply, or
c. energy consumers to include a given proportion of energy from renewable sources in their consumption.

These obligations are seen as proven drivers for:

- Energy security[62]
- Market integration

[59]Supra Note, 49. Directive 2009/28/EC.

[60]Such commitments include the amended Montreal Protocol provisions with the aim of global phase-down of highly global warming hydrofluorocarbons (HFCs) and in the International Maritime Organization (IMO) to an agreement towards an emission reduction strategy for the international shipping sector.

[61]Directive 2009/28/EC, Article 2(l).

[62]It is reported that using more renewables resulted in a €16 billion saving in fossil fuel imports in 2015, and this is projected to rise to €58 billion in 2030. EU Commission, Renewable Energy—Progress Reports, https://ec.europa.eu/energy/en/topics/renewable-energy/progress-reports.

- Energy efficiency
- Decarbonisation
- Innovation[63]

The RED stipulates that by 2020 at least 20% of EU general energy use and at least 10% of transport fuels must come from renewable energy sources.[64] It is worth noting that in 2016, following an agreement at the European Council, the EU Commission published a proposal for a revised renewable energy directive in order to make the EU a global leader in renewable energy and ensure that the target of at least 27% of renewables in the final energy consumption of the EU is met by 2030. In other words, the recast RED will provide a 2030 framework. In addition, under the recast RED, by January 2021 MS shall create contact points for submitting applications and licencing in the area of building and operation of facilities using renewable energy sources. The recast RED also proposes the extension of *guarantees of origin* to renewables whereby green energy sources can be properly certified and identified. These new provisions are an indication that at the supranational level the EU is not complacent about existing targets; on the contrary, it is continually seeking to improve energy security by increasing the energy use efficiency of its MS.

As is the case with other legal commitments, the EU Commission biennially monitors the specific national renewable energy targets that are expressed as a percentage of gross final consumption of energy for all EMS, taking into account their unique initial capacity and overall potential for renewables. Accordingly, while Malta has a target of 10%, Ireland has a target of 16% and Sweden has a target of 49% by 2020.[65] Article 4 of the RED requires MS to report their progress every two years.

[63]The EU has 30% of global patents in renewables and the turnover of the renewables industry in 2014 was €144 billion. Ibid.

[64]The preamble, para. 13 and Article 3 (4), Directive 2009/28/EC.

[65]Overall 2020 renewable energy targets of each EU Member State can be found here: https://ec.europa.eu/energy/en/topics/renewable-energy/national-action-plans.

The RED also encourages cooperation among countries within and outside the EU to meet their respective renewable energy targets through, *inter alia,* statistical transfers of renewable energy,[66] joint renewable energy projects[67] and joint renewable energy support schemes.[68] Statistical transfers refer to cases when a MS transfers its renewable energy to another MS without reporting this data. Thus, the statistics do not reflect the true use of renewable energy across the EU, which has an impact on the accuracy of statistics and on achieving certain targets. As for an example of a joint project/scheme, a High-Level Group was set up for energy cooperation between the Northern Sea countries to integrate offshore wind and enhance interconnections.[69]

Since the RED came into force, the ongoing reporting has revealed that emission levels have reduced by 22% compared to the levels in 1990,[70] which is well short of the binding target of at least 40% reduction in greenhouse gas emissions by 2030 compared to 1990, and the share of renewables reached 16% of the gross final energy consumption of the EU. In order to meet the targets for 2030, the EU needs to increase its final renewable electricity consumption by 31% compared to the 2016 figure.[71]

Despite the limited achievements in reduction of emissions, it is reported that the output of environmental goods and services per unit of Gross Domestic Product (GDP) displayed at least 50% growth over the last decade and the employment linked to this 'green economy' also grew to more than 4 million full-time equivalents.[72] Waste-to-Energy and Strategic Energy Technology[73] initiatives can also be attributed to

[66]As expressed by Article 6 of RED.

[67]As expressed by Articles 7 and 8 of RED.

[68]As expressed by Article 11 of RED.

[69]EU Commission, Energy, North Seas countries agree on closer energy cooperation, 6 June 2016, https://ec.europa.eu/energy/en/news/north-seas-countries-agree-closer-energy-cooperation.

[70]Communication from the Commission to the European Parliament, the Council, the European Economic and Social Committee, the Committee of the Regions and the European Investment Bank—Second Report on the State of the Energy Union, COM(2017) 53 final, https://ec.europa.eu/commission/sites/beta-political/files/2nd-report-state-energy-union_en.pdf.

[71]Ibid.

[72]Ibid.

[73]European Commission, Strategic Energy Technology Plan, https://ec.europa.eu/energy/en/topics/technology-and-innovation/strategic-energy-technology-plan.

the RED.[74] Thanks to the impetus from the RED, a substantial amount of investment in renewables has been made. For instance, more than 20% of investment supported by the European Fund for Strategic Investments was related to the energy field, through which investment for interconnectors was launched. This investment includes the Central East South Europe Gas Connectivity group, the mandate of which is envisaged to extend to electricity, renewables and energy efficiency.[75] To meet growing demand, the EU needs to increase its renewable energy capacity by 90% (Banja and Jégard 2017, p. 2). This has been a significant challenge for the EU because economic stagnation has continued since 2008, while the EU registered a nominal increase in the volume of trade and related revenues only in 2018.[76] In addition, the EU has also lost its dominant position in its share of renewable energy jobs per capita in the labour force: although in 2015 the EU had ranked second (after Brazil), in the 2017 report the EU ranked fifth, while Japan, the United States and China were ahead.[77]

One of the most innovative aspects of the RED has been the interpretation of the *renewable energy obligation*, whereby consumers are conferred tangible benefits via not only investment and improved technologies but also via new market design, which empowers consumers to form renewable energy cooperatives, generate energy and contribute to the energy supply mix.[78] The recast RED is more specific and obligatory

---

[74]Communication from the Commission to the European Parliament, the Council, the European Economic and Social Committee and the Committee of the Regions—The role of waste-to-energy in the circular economy, COM(2017) 34 final, http://ec.europa.eu/environment/waste/waste-to-energy.pdf.

[75]Supra note, 83.

[76]Eurostat, 'Euro area international trade in goods surplus €3.3 bn', 46/2018 19 March 2018, http://trade.ec.europa.eu/doclib/docs/2013/december/tradoc_151969.03.2018.pdf.

[77]European Environment Agency, Renewable energy in Europe 2017—EEA Report No. 3/2017.

[78]The following proposed laws stipulate such provisions to be put in place: Proposal for a Directive of the European Parliament and of the Council on common rules for the internal market in electricity (recast), COM(2016) 864 (Electricity Directive) https://eur-lex.europa.eu/legal-content/EN/TXT/?uri=CELEX%3A52016PC0864; and Proposal for a Regulation of the European Parliament and of the Council on the internal market for electricity—(recast), COM(2016) 861 final/2; https://ec.europa.eu/energy/sites/ener/files/documents/1_en_act_part1_v9.pdf.

in this regard; it states that MS shall ensure that self-consumers who generate electricity can consume it without undue restrictions and can sell the excess of produced energy to the grid, and renewable energy cooperatives are not to be discriminated against in accessing energy markets. This is a significant opportunity for individual households, small energy producers and energy cooperatives in the EU. However, the EU will only be able to maximise the benefits if there are effective interconnections among them and improved transfer and storage capacities.[79]

Shortcomings of the Renewable Energy Directive

Despite all these positive features and results of the RED, a number of contentious issues and shortcomings remain. For example, the size of renewable energy potential in the EU is still unknown and a comparison of capacity in renewables among the EU MS is yet to be conducted.[80] Despite significant investment in renewables,[81] overall consumption of fossil fuels remains high in the EU. Furthermore, while in 2016, electricity in the EU from renewable sources increased to almost 40% of total capacity, the volume of gas consumption increased twice as fast as that of renewables.[82] This continuing reliance on fossil fuels hinders efficiency and renewable energy production targets.[83] The patchwork of efficiency achievements is also compounded by the

---

[79]Namely, energy security, market integration, energy efficiency, decarbonisation; and innovation.

[80]There are statistics which compares overall EU capacity with other regions in the world. See for example, International Renewable Energy Agency, Global Overview of the Renewable Energy Installed Capacity and Electricity Generated, 2016, http://resourceirena.irena.org/gateway/dashboard/?topic=4&subTopic=17.

[81]Also, note that there is no comparative data in terms of regional investment in renewables across the EU.

[82]Banja and Jégard indicated that during 2005–2015, "the overall renewable energy share increased by an annual average of 0.8 percentage points. In the same period, final renewable energy consumption increased by an average of 7.8 Mtoe per year", p. 4. The decreasing reliance on nuclear energy has also contributed to this trend.

[83]For example, Germany has the largest wind-power capacity, with 44% of all new EU installations. Spain, the UK and France also have large shares compared to remaining EU Member States which have nominal capacity in terms of wind power generation.

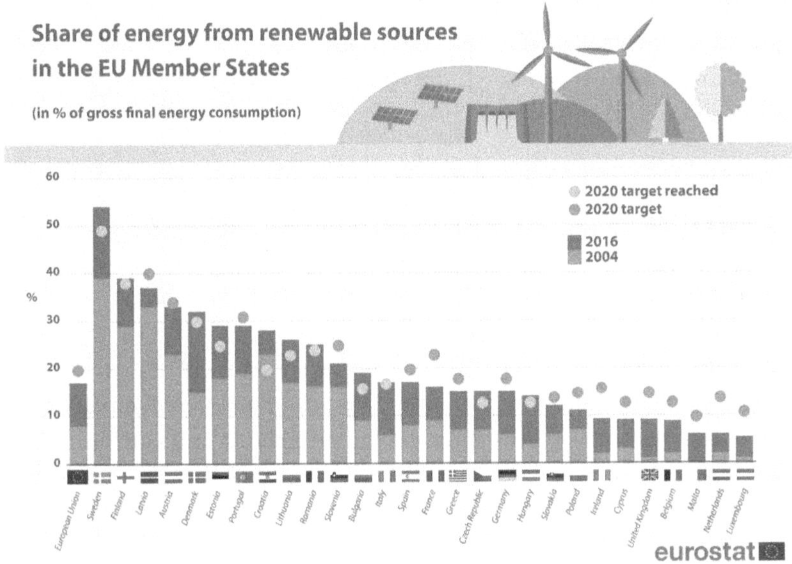

Fig. 7.3 Share of energy from renewable sources in the EU 2004–2016 (*Source* Eurostat, Share of Renewable Energy Sources 2018, http://ec.europa.eu/eurostat/statistics-explained/index.php?title=File:Figure_1-Share_of_energy_from_renewable_sources_2004-2016.png)

variety of renewable energy sources available to the MS and their varying ability to utilise them.[84] Therefore, the patterns of energy efficiency development are similar, although achievements are quite different.[85] For example, Denmark, with abundant onshore and offshore winds, has been able to produce nearly 30% of its energy from wind power and generate 11,000 additional jobs in this sector. By contrast, Ireland, with similar demographic characteristics and GDP, generates less than 10% of its energy from renewables, as Fig. 7.3 shows.[86]

[84]EEA Report No. 3/2017 (pp. 16–29) provides a good overview of the level of renewable energy generation from various sources such as wind, solar, hydro and thermal power, etc.

[85]The EU Commission, Progress Reports, https://ec.europa.eu/energy/en/topics/renewable-energy/progress-reports.

[86]Eurostat, Share of energy from renewable sources in the EU Member States, 2004–2016, http://ec.europa.eu/eurostat/statistics-explained/index.php?title=File:Figure_1-Share_of_energy_from_renewable_sources_2004-2016.png.

Despite such mixed capacity and results across the EU, the rate of growth in the EU's renewable electricity sector has been 7% per year, on average.[87] If this growth is sustained, the energy efficiency targets set by the EU may be achieved.

Conclusion

This chapter has identified the key legal and policy instruments created to achieve energy use efficiency in the EU and provided a critical analysis of the priority areas therein. The EU has set out several energy strategies for a European energy union, which aim to ensure more secure, efficient, sustainable, competitive and affordable energy production and consumption.[88] These strategies have enabled the EU to spearhead global investment in energy efficiency and energy transformation. While the EU has succeeded in establishing a common vision and policies for energy efficiency in the EU's Single Market, varying conditions in MS have exerted strong impact on meeting the targets set, thus yielded mixed results. These vastly varying conditions include different starting points, developmental needs, technological and infrastructural capacities (i.e. generation, storage, transfer), resource potential and specific market conditions among the MS.[89]

[87]European Environment Agency, Renewable energy in Europe 2017—EEA Report No. 3/2017, p. 6.

[88]For example, the Roadmap 2011; Communication from the Commission to the European Parliament, The Council, the European Economic and Social Committee and the Committee of the Regions Renewable Energy: a major player in the European energy market—COM/2012/0271 Final; and other legal and policy instruments as outlined in this chapter.

[89]For example, with regard to electric car charging points, the Netherlands leads the way with a network of over 23,000 public charging positions followed by Germany with more than 14,000, France with more than 13,000, the UK around 11,500 and Norway with more than 7600 whereas Bulgaria, Cyprus, Iceland and Lithuania have fewer than 40 charging points. See, European Alternative Fuels Observatory, Electric vehicle charging infrastructure, http://www.eafo.eu/electric-vehicle-charging-infrastructure. Interestingly, the highest shares of renewable energy attained in the EU belong to Sweden (52.6%), followed by Finland (38.7%) and Latvia (38.7%). Luxembourg (4.5%), Malta (4.7%) and the Netherlands (5.5%) which realised the lowest shares! Supra note, 97, EEA Report No. 3/2017, p. 15.

Furthermore, the global economic downturn[90] has had a negative impact on energy efficiency.

These trends indicate that, although the EU has a common vision on energy efficiency and continually reviews its targets, reliance on policy and law is not enough to bring about the desired change. Firstly, substantial investment in efficiency and capacity building (for renewable energy generation) is necessary. Secondly, energy interconnections and transfer hubs and networks within and outside the EU are necessary for the MS to benefit from each other's conditions, such as varying supply and demand, and advantages.

With the eventual departure of the UK from the EU (Brexit), such patterns may be compounded further as relevant legal provisions, like the EED and RED, may no longer apply in the UK. This could have consequences on both regulatory standards, such as Guarantees of Origin,[91] and overall achievements in the EU. Nevertheless, the EU is interested in making progress with its Roadmap and an ambitious agenda for the green economy. Consequently, renewable energy continues to present a great opportunity for the EU and its 27 member states, with proven prospects of boosting their economic growth, creating new jobs requiring specific skills and value chains and attracting investment, while delivering much needed social and environmental benefits for all.

References

Alhajji, A. F. (2008). *What is energy security? Middle East economic survey, 5/5.*

Alikhanzadeh, A. H., & Taylor, G. A. (2014). *Future European electricity highways: A comparative pan-European and GB transmission system perspective.*

---

[90]For example, economic crisis and increased competition resulted in Europe's investment share to decline from 46% in 2005 to 17% in 2015.

[91]Article 15(2) of the EED requires Member States to ensure that a guarantee of origin is issued in response to a request from a producer of electricity from renewable energy sources. In the event that the EED seizes to apply in the UK, guarantees of origin that have been issued by designated bodies in the UK in accordance with Article 15(2) of Directive 2009/28/EC will no longer be recognised by the EU-27 Member States as of the withdrawal date.

International Conference on Power System Technology, 20–22 October. http://ieeexplore.ieee.org/abstract/document/6993994/.
Banja, M., & Jégard, M. (2017). *Renewable technologies in the EU electricity sector: Trends and projections*. Publication Office of the EU. https://ec.europa.eu/jrc/en/publication/eur-scientific-and-technical-research-reports/renewable-technologies-euelectricity-sector-trends-and-projections-analysis-framework-eu.
Barton, B., Redgwell, C., Ronne, A., & Zillman, D. (2004). Introduction. In B. Barton, C. Redgwell, A. Ronne, & D. Zillman, *Energy security—Managing risk in as dynamic legal and regulatory environment*. Oxford: Oxford University Press.
Belyi, A., Nappert, S., & Pogoretskyy, V. (2011). Modernising the energy charter process? The energy charter conference road map and the Russian draft convention on energy security. *Journal of Energy & Natural Resources Law, 29*(3), 383–399.
Bloomberg New Energy Finance. (2014). 2030 Market Outlook.
Brilhante, O., & Skinner, J. (2014). *European experiences of building codes for promoting sustainable housing*, Report No. 1. The Netherlands: Institute for Housing and Urban Development Studies (IHS) Rotterdam.
Cameron, P. (2007). *Competition in energy markets—Law and regulation in the European Union*. Oxford: Oxford University Press.
Campus Oil. (1984). Case 72/83—Judgment of the Court of 10 July.
Commission Priority, Energy Union: Making energy more secure, affordable and sustainable. Available at: http://ec.europa.eu/priorities/energy-union/index_en.htm.
Communication from the Commission to the European Parliament, the Council, the European Economic and Social Committee and the Committee of the Regions Roadmap to a Resource Efficient Europe. (2011). COM/2011/0571 final. Available at: http://eur-lex.europa.eu/legal-content/EN/TXT/?uri=CELEX:52011DC0571.
Communication from the Commission to the European Parliament, the Council, the European Economic and Social Committee and the Committee of the Regions Closing the Loop—An EU action plan for the Circular Economy. (2015). COM/2015/0614 final. Available at: http://eur-lex.europa.eu/legal-content/EN/TXT/?uri=CELEX:52015DC0614.
Communication from the Commission to the European Parliament, the Council, the European Economic and Social Committee and the Committee of the Regions—The role of waste-to-energy in the circular economy. (2017). COM(2017) 34 final. Available at: http://ec.europa.eu/environment/waste/waste-to-energy.pdf.

Communication from the Commission to the European Parliament, the Council, the European Economic and Social Committee, the Committee of the Regions and the European Investment Bank—Second Report on the State of the Energy Union. (2017). COM(2017) 53 final. Available at: https://ec.europa.eu/commission/sites/beta-political/files/2nd-report-state-energy-union_en.pdf.

Communication from the Commission to the European Parliament, the Council, the European Economic and Social Committee, Energy Roadmap 2050. Available at: http://eur-lex.europa.eu/legal-content/EN/ALL/?uri=CELEX%3A52011DC0885.

Cottier, T., Malumfashi, G., Matteotti-Berkutova, S., Nartova, O., de Sepibus, J., & Bigdeli, S. (2010). *Energy in WTO law and policy*. Available at: http://www.wto.org/english/res_e/publications_e/wtr10_forum_e/wtr10_7may10_e.pdf.

Council Directive. (2004). 2004/67/EC of 26 April concerning measures to safeguard security of natural gas supply, OJ L 127, 29/04/2004, Preamble, para.13.

Directive 2000/60/EC of the European Parliament and of the Council of 23 October 2000 establishing a framework for Community action in the field of water policy. Available at: http://eur-lex.europa.eu/legal-content/EN/TXT/?uri=CELEX:32000L0060.

Directive 2009/28/EC of the European Parliament and of the Council of 23 April 2009 on the promotion of the use of energy from renewable sources and amending and subsequently repealing Directives 2001/77/EC and 2003/30/EC. Available at: http://eur-lex.europa.eu/legal-content/EN/ALL/?uri=celex%3A32009L0028.

Directive 2009/72/EC concerning common rules for the internal market in electricity and repealing Directive 2003/54/EC.

Directive 2009/73/EC concerning common rules for the internal market in natural gas and repealing Directive 2003/55/EC.

Directive 2009/125/EC of the European Parliament and of the Council of 21 October 2009 establishing a framework for the setting of eco-design requirements for energy-related products. Available at: http://eur-lex.europa.eu/legal-content/EN/TXT/?uri=CELEX:32009L0125.

Directive 2010/31/EU of the European Parliament and of the Council of 19 May 2010 on the energy performance of buildings. Available at: http://eur-lex.europa.eu/legal-content/EN/ALL/?uri=celex:32010L0031.

Directive 2011/92/EU of the European Parliament and of the Council of 13 December 2011 on the assessment of the effects of certain public and private projects on the environment Text. Available at: http://eur-lex.europa.eu/legal-content/EN/TXT/?uri=celex%3A32011L0092.

Directive 2012/27/EU of the European Parliament and of the Council of 25 October 2012 on energy efficiency, amending Directives 2009/125/EC and 2010/30/EU and repealing Directives 2004/8/EC and 2006/32/EC. Available at: http://eur-lex.europa.eu/LexUriServ/LexUriServ.do?uri=OJ:L:2012:315:0001:0056:en:PDF.

European Alternative Fuels Observatory, Electric vehicle charging infrastructure. Available at: http://www.eafo.eu/electric-vehicle-charging-infrastructure.

European Environment Agency. (2017). *Renewable energy in Europe*. EEA Report No. 3/2017.

European Commission. (2006). *A European strategy for sustainable, competitive and secure energy* (Green Paper) COM 105 Final.

European Network of Transmission System Operators for Gas, Security of Gas Supply. (2017). Available at: https://www.entsog.eu/publications/security-of-gas-supply.

European Parliament Resolution on a Resource-Efficient Europe. (2012). 2011/2068(INI) 24 May.

EU Commission. (2000). *Towards a European strategy for the security of energy supply* (Green Paper). COM(2000) 769 Final.

EU Commission, A fully integrated internal energy market. Available at: https://ec.europa.eu/commission/priorities/energy-union-and-climate/fully-integrated-internal-energy-market_en.

EU Commission, Clean Energy for all Europeans—Good practice in energy efficiency. (2016). https://ec.europa.eu/energy/sites/ener/files/documents/good_practice_in_ee_-web.pdf.

EU Commission, Energy, North Seas countries agree on closer energy cooperation. (2016, June 6). Available at: https://ec.europa.eu/energy/en/news/north-seas-countries-agree-closer-energy-cooperation.

EU Commission, EU Energy Policy Data. (2007). SEC (2007), 12.

EU Commission, Energy Efficiency. Available at: https://ec.europa.eu/energy/en/topics/energy-efficiency.

EU Commission, Financing Energy efficiency. Available at: https://ec.europa.eu/energy/en/topics/energy-efficiency/financing-energy-efficiency.

EU Commission, Internal energy market: Commission refers Bulgaria, Estonia and the United Kingdom to Court for failing to fully transpose EU rules. (2013). Press Release Database, 24 January. Available at: http://europa.eu/rapid/press-release_IP-13-42_en.htm.

EU Commission, GPP—A Collection of good practices. Available at: http://ec.europa.eu/environment/gpp/pdf/GPP_Good_Practices_Brochure.pdf.

EU Commission, Market Observatory for Energy, *Key Figures*. (2011). http://ec.europa.eu/energy/observatory/eu_27_info/doc/key_figures.pdf.

EU Commission—Press Release, Towards Energy Union: Sustainable energy security package, Brussels. (2016, February 16). Available at: http://europa.eu/rapid/press-release_IP-16-307_en.htm.

EU Commission, Progress Reports. Available at: https://ec.europa.eu/energy/en/topics/renewable-energy/progress-reports.

EU Commission, Renewable Energy—Progress Reports. Available at: https://ec.europa.eu/energy/en/topics/renewable-energy/progress-reports.

EU Commission, Single Market Act II—Together for New Growth. (2012). COM(2012) 573 Final.

EU Commission, Strategic Energy Technology Plan. Available at: https://ec.europa.eu/energy/en/topics/technology-and-innovation/strategic-energy-technology-plan.

Eurostat, Share of Energy from Renewable Sources in the EU Member States, 2004–2016. (2018). Available at: http://ec.europa.eu/eurostat/statistics-explained/index.php?title=File:Figure_1-Share_of_energy_from_renewable_sources_2004-2016.png.

Eurostat, Euro Area International Trade in Goods Surplus €3.3 bn. (2018). 46/2018 19 March, available at: http://trade.ec.europa.eu/doclib/docs/2013/december/tradoc_151969.03.2018.pdf.

Flynn, C. (2006). Russian roulette: The ECT, transit and Western European energy security. *Oil Gas & Energy Law, 4*. https://www.ogel.org/article.asp?key=2288.

Grimwood, T. (2018, February 6). *Hinkley and Sizewell: Power in numbers.* Utility week. Available at: https://utilityweek.co.uk/hinkley-sizewell-power-numbers/.

International Energy Agency. (2014). World Energy Investment Outlook.

International Energy Agency, Energy Efficiency. (2017). Available at: https://www.iea.org/publications/freepublications/publication/Energy_Efficiency_2017.pdf.

International Energy Outlook. (2004). Available at: http://tonto.eia.doe.gov/FTPROOT/forecasting/0484%282004%29.pdf.

International Energy Outlook. (2005). Available at: http://tonto.eia.doe.gov/FTPROOT/forecasting/0484%282006%29.pdf.

International Renewable Energy Agency, Global Overview of the Renewable Energy Installed Capacity and Electricity Generated. (2016). http://resourceirena.irena.org/gateway/dashboard/?topic=4&subTopic=17.

Kuzemko, C. (2014). Ideas, power and change: Explaining EU–Russia energy relations. *Journal of European Public Policy, 21*(1), 58–75.

Maican, O.-H. (2009). Some legal aspects of energy security in the relations between EU and Russia. *Romanian Journal of European Affairs, 9*, 4.
Plebalgs, A. (2009). EU-Russia energy relations: Common goals and concerns. *Oil Gas & Energy Law, 2*. https://www.ogel.org/article.asp?key=2867.
Pomerantz, D. (2017, September 12). Size matters: The next big thing in wind turbines. GE reports—Wind power. Available at: https://www.ge.com/reports/size-matters-next-big-thing-wind-turbines/.
Preamble of the EU Energy Efficiency Directive 2012, Directive 2012/27/EU of the European Parliament and of the Council of 25 October 2012 on energy efficiency, amending Directives 2009/125/EC and 2010/30/EU and repealing Directives 2004/8/EC and 2006/32/EC.
Proposal for a Directive of the European Parliament and of the Council on the promotion of the use of energy from renewable sources (recast). (2016). COM/2016/0767 final/2—2016/0382 (COD), section 1.1. Available at: http://eur-lex.europa.eu/legal-content/EN/TXT/?uri=CELEX:52016PC0767R%2801%29#footnoteref2.
Proposal for a Directive of the European Parliament and of the Council on common rules for the internal market in electricity (recast). (2016). COM(2016) 864 (Electricity Directive). Available at: https://eur-lex.europa.eu/legal-content/EN/TXT/?uri=CELEX%3A52016PC0864.
Proposal for a Regulation of the European Parliament and of the Council on the internal market for electricity (recast). (2016). COM(2016) 861 final/2: https://ec.europa.eu/energy/sites/ener/files/documents/1_en_act_part1_v9.pdf.
Regulation (EC) No. 715/2009 on conditions for access to the natural gas transmission networks and repealing Regulation (EC) No. 1775/2005.
Regulation (EC) No. 714/2009 on conditions for access to the network for cross-border exchanges in electricity and repealing Regulation (EC) No. 1228/2003.
Regulation (EC) No. 713/2009 of the European Parliament and of the Council of 13 July 2009 establishing an Agency for the Cooperation of Energy Regulators.
Regulation (EU) No. 347/2013 on guidelines for trans-European energy infrastructure and repealing Decision No. 1364/2006/EC and amending Regulations (EC) No. 713/2009, (EC) No. 714/2009 and (EC) No. 715/2009.
Regulation (EU) No. 994/2010 of the European Parliament and of the Council of 20 October 2010 concerning measures to safeguard security of gas supply and repealing Council Directive 2004/67/EC. Note that Regulation No. 994/2010 was repealed by Regulation (EU) 2017/1938 of the European

Parliament and of the Council of 25 October 2017 concerning measures to safeguard the security of gas supply: https://eur-lex.europa.eu/legal-content/en/TXT/?uri=CELEX:32017R1938.

Report from the Commission to the European Parliament and the Council, Assessment of the progress made by Member States towards the national energy efficiency targets for 2020 and towards the implementation of the Energy Efficiency Directive 2012/27/EU as required by Article 24(3) of Energy Efficiency Directive 2012/27/EU, COM(2015) 574 final. https://ec.europa.eu/energy/sites/ener/files/documents/2a_EE%20progress%20report%20-%20CSWD%20part%201.pdf.

Report from the Commission to the European Parliament and the Council 2017 assessment of the progress made by Member States towards the national energy efficiency targets for 2020 and towards the implementation of the Energy Efficiency Directive as required by Article 24(3) of the Energy Efficiency Directive 2012/27/EU. http://eur-lex.europa.eu/legal-content/EN/TXT/?qid=1511978095545&uri=COM:2017:687:FIN.

Ricardo Energy & Environment. *Study evaluating progress in the implementation of Article 7 of the Energy Efficiency Directive*. Report for DG energy, ENER.C3.dir(2014)3156530. Available at: https://ec.europa.eu/energy/sites/ener/files/documents/final_report_evaluation_on_implementation_art._7_eed.pdf.

Roadmap 2011. (2011). Communication from the Commission to the European Parliament, the council, the European Economic and Social Committee and the Committee of the regions renewable energy: A major player in the European energy market—COM/2012/0271 Final.

Rosenow, J., Leguijt, C., Pató, Z., Eyre, N., & Fawcett, T. (2016). An ex-ante evaluation of the EU Energy Efficiency Directive—Article 7. *Economics of Energy and Environmental Policy, 5*(2), 45–63.

Saga, B. (1995). Introduction of competition in gas markets: Effects on contract structures and security of supply. In L. Hancher (Ed.), *The European energy market: Reconciling competition and security of supply*. Postanschrift: Bundesanzeiger.

Selivestrov, S. (2009). Energy security of Russia and the EU: Current legal problems. *Oil Gas & Energy Law, 2*. http://www.ogel.org/article.asp?key=2896.

Turksen, U. (2018). *EU energy relations with Russia: Solidarity and the rule of law*. Abingdon: Routledge.

United Nations, Climate Change, Kyoto Protocol. Available at: http://unfccc.int/kyoto_protocol/items/2830.php.

UN Country Classifications. Available at: http://www.un.org/en/development/desa/policy/wesp/wesp_current/2014wesp_country_classification.pdf.
Wagnsson, C. (2010). Divided power Europe: Normative agencies among the EU "big three". *Journal of European Public Policy, 17*(8), 1089–1105.
Youngs, R. (2010). *The EU's role in world politics: A retreat from liberal internationalism*. Abingdon: Routledge.

8

The Success of the South West of the UK in Renewable Energy Generation: Benefits, Challenges and Implications for Other Regions

Alison Ashby

Introduction

The UK government has committed to moving towards a low carbon economy, evidenced by strong policies towards the promotion of renewable energy (Wood and Dow 2011). The UK is fortunate to have a wealth of energy resources, but has relied heavily on the use of coal, oil and gas supplies for homes, businesses and transport. The depletion of domestic fossil fuel reserves, combined with growth in global demand for energy, puts the UK's energy security at risk. Security of supply, fossil fuel depletion and climate change are the three main drivers for a low carbon economy (Wood and Dow 2011). The UK government's drive to increase the proportion of energy obtained from renewable sources will not only increase the nation's energy security and reduce its reliance on fossil fuels and external supply but will also provide opportunities for investment in new industries and new technology.

A. Ashby (✉)
University of Plymouth, Plymouth, Devon, UK
e-mail: alison.ashby@plymouth.ac.uk

N. Mouraviev and A. Koulouri (eds.), *Energy Security*,
https://doi.org/10.1007/978-3-030-01033-1_8

The south west of the UK is one of nine official regions of England, and is the largest in area, covering 9200 square miles and with 5 million inhabitants. It consists of the counties of Gloucestershire, Bristol, Wiltshire, Somerset, Dorset, Devon and Cornwall, and the Isles of Scilly. The south west is widely recognised as a regional leader in the extent and range of its use of renewable energy and in resource efficiency. In this region, solar, wind, wave and biomass energy are well developed for both commercial and domestic use. The region has 2.5 times more solar capacity than the regional average and 1.5 times more than its closest competitor (www.gov.uk/government/statistics/regional-renewable-statistics). Due to its favourable climate conditions and successful promotion of renewable energy, the region has more than a fifth of all UK renewables' projects that are being implemented by homeowners, landowners and businesses.

The chapter discusses arrangements in local and community-scale renewables that bring benefits to the region in the form of greater energy security, job creation and higher incomes. It highlights the current debate in the literature in which there is a growing recognition of the importance of implementation of identified actions in the field of renewable energy at the local and regional levels, rather than at the national level, where energy policy decision-making typically takes place. This chapter therefore investigates how and why this UK region has, to date, been so successful in adopting the different forms of renewable energy, identifies their benefits and challenges, and evaluates the implications these local achievements have for other regions and the UK as a whole.

The chapter undertakes a review of green energy practice, specifically within the counties of Devon and Cornwall where renewable energy use is the highest, and draws on data from county councils, government, regional energy reports and independent sources. It contributes to the edited collection by providing insights into how regional characteristics enable the extensive adoption of multiple forms of renewable energy and highlights the distinctive features of a collaborative approach employed by the industry, local authorities and regional agencies that support these activities.

Defining Renewable Energy

Renewable energy is a subset of sustainable energy and relates to the capture of energy from existing flows of energy, from ongoing natural processes. Modern interest in renewable energy development is linked to concerns about the exhaustion of fossil fuels and environmental, social and political risks of the extensive use of fossil fuels and nuclear energy (Ellabban et al. 2014). Renewable energy is commonly defined as energy from a resource that is replaced by a natural process, at a rate equal to or faster than the rate at which the resource is being consumed. While theoretically renewable on a very long time-scale, fossil fuels are exploited at rates that may deplete these resources in the near future and are therefore, not considered renewable. Renewable energy resources may be used directly or used to create other more convenient forms of energy. Examples of direct use are solar ovens, geothermal heating, and windmills. Examples of indirect use which require energy harvesting, are electricity generation through wind turbines or photovoltaic cells (PV cells), or production of fuels, such as biogas, from anaerobic digestion (Ellabban et al. 2014).

Electricity generation from renewable energy sources is supported in many countries, and fixed feed-in tariffs (FITs) have been a dominant policy instrument utilised for the support of renewable energy. FIT applies to the small-scale generation of electricity using eligible renewable technologies and the energy producer is paid a set financial amount even for the energy they consume themselves. Due to government support, since 2010 there was a surge in investment in renewable energy schemes, in particular, solar PV, which together with the progressive fall in the costs of technology and installation, led to high levels of adoption of this form of renewable energy across the UK. However, this rapid and significant growth in solar PV implementation and the widening variety of incentives offered to small-scale producers of electricity have caused complexity in the energy market and a skewed focus on specific renewable technologies. There is an increasing pressure to improve the market integration of renewable energy, i.e. to make multiple forms of renewable energy easily available to all customers without relying on financial incentives. As a result, many countries have re-evaluated the use of FIT schemes (Kitzing and Weber 2015). In the UK, cuts to financial support, which have been implemented since 2015, has slowed down investments in renewables (Regensw 2016).

UK National Energy Policy

The 2009 Renewable Energy Directive set a target for the UK to achieve 15% of its energy consumption from renewable sources by 2020, compared to only 1.5% in 2005 (www.gov.uk/government/uploads/system/uploads/attachment_data/file/47871/25-nat-ren-energy-action-plan.pdf). As Fig. 8.1 indicates, there was only a small increase in renewable energy use ahead of this directive, and consequently, a greater level of deployment is required to meet the specified target. The National

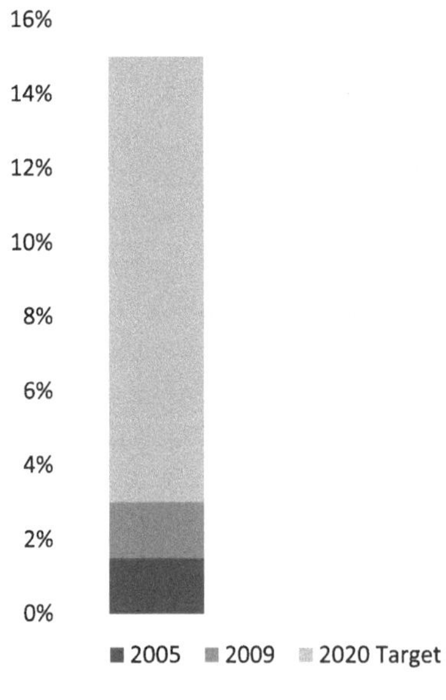

Fig. 8.1 Proportion of energy consumed from renewable sources in the UK, 2005–2020 (*Source* Adapted by the author from www.gov.uk/government/uploads/system/uploads/attachment_data/file/622815/Renewable_energy_in_2016.pdf)

Renewable Energy Action Plan provides details on a set of measures that would enable the UK to meet its 2020 target, but it also aims to secure energy supplies beyond 2020 and provide a sound framework for business to develop in new industries and provide jobs in the energy sector. This requires exploiting the UK's renewable energy potential to the greatest extent possible.

The history of energy production in the UK has been based on the use of its natural resources of fossil fuels, which means that it has not been active in its exploitation of available renewable resources. Compared to many other EU member states, the UK is starting from a very low level of renewable energy consumption and therefore, meeting the 2020 target presented in Fig. 8.1 is a challenging task. The independent UK Committee on Climate Change was commissioned to review the renewables target and provide advice on increasing the level of ambition. The Committee aims to make an Annual Energy Statement to the UK Parliament to set strategic energy policy and guide investment in all forms of energy including renewables.

The use of renewable energy sources is seen as a key element in the UK's energy policy, reducing the dependence on fuel imported from non-EU countries, reducing emissions from fossil fuel sources and decoupling energy costs from oil prices. The government Directive 2009/28/EC on the promotion of the use of energy from renewable sources established accounting criteria for the proposed 2020 targets for renewable energy (www.gov.uk/government/uploads/system/uploads/attachment_data/file/622815/Renewable_energy_in_2016.pdf). Figure 8.2 illustrates the UK's progress against its interim targets since 2004, which it has consistently exceeded, towards the 2020 target of 15% of energy from renewables. It currently meets 5% of its total energy demand from renewables (www.gov.uk/government/uploads/system/uploads/attachment_data/file/622815/Renewable_energy_in_2016.pdf) and performs particularly well in its generation of electricity, with 19% of total electrical demand coming from renewable sources.

One of the main support mechanisms for large-scale renewable energy projects to achieve these challenging targets is the Renewables Obligation (RO), which came into effect in 2002 and is administered by the Office of Gas and Electricity Markets (Ofgem). It requires

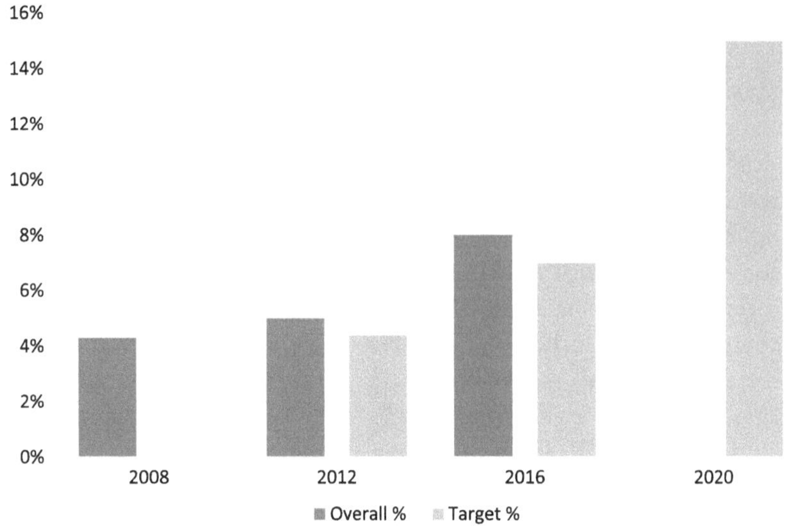

Fig. 8.2 Progress against renewable energy directive and UK targets (*Source* Adapted by the author from www.gov.uk/government/uploads/system/uploads/attachment_data/file/622815/Renewable_energy_in_2016.pdf)

licensed electricity suppliers to buy an increasing percentage of their supply from eligible renewable resources (www.ofgem.gov.uk/environmental-programmes/ro/about-ro). The RO allows the cost burden of this enforced change in supply to be offset by receiving Renewables Obligation Certificates (ROCs) (Hain et al. 2005), i.e. green certificates which are issued by Ofgem to accredited renewable electricity generators (www.ofgem.gov.uk/environmental-programmes/ro/about-ro).

One ROC is typically issued for each megawatt-hour (MWh) of eligible renewable output, although there are differences between certain kinds of renewable technologies. Where suppliers do not present a sufficient number of ROCs to meet their obligation in the one-year reporting period, they must pay an equivalent amount into a buy-out fund. Once the administration cost of the scheme is recovered, the rest of the fund is distributed back to suppliers in proportion to the number of ROCs they produced in relation to their individual obligation. The expected supplier obligation is 100.7 million ROCs, but in 2016–2017 only 86.2 million were issued. When combined with electricity generated by FIT installations, renewable generation under the RO was equivalent to 24.8% of total UK electricity supply (www.ofgem.gov.uk/system/files/docs/2018/03/ro_annual_report.pdf).

Forms of Renewable Energy

There are multiple forms of renewable energy sources that the UK and other countries are employing to meet their energy targets, reduce reliance on fossil fuels and address issues, such as climate change. The extent of their applicability and use depends on the availability of natural resources and technological development within the country and/or region (Regensw 2016). Table 8.1 illustrates the different forms of renewable energy currently employed in the UK, and the percentage delivered by each in Gigawatts per hour (GWh). Biomass, wind power and solar PV are currently the dominant forms of renewable energy, and this reflects the extent to which the technology has been developed to support them as well as their more natural abundance in comparison to other forms.

Renewable energy sources are those resources that can be used to produce energy again and again (Panwar and Kaushik 2011), and a number of sources are available as illustrated in Table 8.2. Solar thermal energy

Table 8.1 Percentage of renewable energy by technology in the UK

Renewable energy technology	% use (GWh)
Anaerobic digestion	3.5
Biomass	26
Energy from waste (EfW)	11
Heat pumps	1.5
Hydro	0.5
Landfill gas	6
Offshore wind	22
Onshore wind	9
Sewage gas	2
Solar PV	18
Solar thermal	0.5

Source Adapted by the author from Regensw (2016)

Table 8.2 Key renewable energy sources and their form of usage

Renewable energy source	Form of usage
Direct solar	Photovoltaic, thermal power generation, water heaters
Wind	Power generation, wind generators, windmills, water pumps
Wave	Various designs
Tidal	Barrage, tidal stream
Hydropower	Power generation
Biomass	Heat and power generation, pyrolysis, gasification, digestion
Geothermal	Urban heating, power generation, hydrothermal, hot dry rock

Source Adapted by the author from Panwar and Kaushik (2011)

is the most abundant and is available in direct and indirect forms. Solar energy generation involves the use of the sun's energy to provide hot water via solar thermal systems or electricity via solar photovoltaic (PV) and concentrating solar power (CSP) systems. The basic building block of PV systems is the PV cell, a semiconductor device that converts solar energy into direct-current electricity. PV systems are highly modular, in that modules can be linked together to provide power ranging from a few watts to tens of megawatts. Solar heating and cooling technologies collect thermal energy from the sun and use this heat to provide hot water, space heating, cooling, and pool heating for residential, commercial, and industrial applications (Ellabban et al. 2014).

Among renewable energy technologies applied to electricity generation, wind energy ranks second only to hydroelectric in terms of installed capacity. Wind energy for electricity production today is a mature, competitive and virtually pollution-free technology. Wind technology converts the energy available in wind to electricity or mechanical power through the use of wind turbines. The first wind turbines for electricity generation were developed at the beginning of the twentieth century and the technology has gradually improved since the early 1970s. By the end of the 1990s, wind energy re-emerged as one of the most important sustainable energy resources (Ellabban et al. 2014).

Biomass energy is the conversion of biomass into useful forms of energy such as heat, electricity and liquid fuels. Biomass for bioenergy comes either directly from the land, such as from dedicated energy crops, or from residues generated in the processing of crops for food or other products. It is a renewable and sustainable form of energy but shares many characteristics with fossil fuels. While biomass can be directly burned to obtain energy, it can also serve as a feedstock to be converted to various liquid or gas fuels (biofuels). Biofuels can be transported and stored, and allow for heat and power generation on demand, which contrasts with intermittent renewable energy sources, such as wind. As a result, biomass is expected to play a major role in future energy scenarios and there is an emerging strategy to develop bio-refinery and biotransformation technologies to convert biomass feedstock into clean energy fuels. However, there are some significant barriers to this strategy: biomass fuels have low energy densities, and collection and transportation can be cost prohibitive. Using biomass to generate electricity is technologically well established, but the price paid for electricity seldom offsets the full cost of the biomass fuel. Bioenergy fuels are also intensive in the use of inputs, which include land, water, crops and fossil energy (Ellabban et al. 2014).

The Regional Perspective

While the world is increasingly turning towards the use of renewable energy sources due to resource constraints and climate change issues, the local, regional and national contexts for innovation and sustainable development are fundamentally different. This means that the stakeholders involved respond differently to national and international directions (Rygg 2014). It is recognised in both research and practice that there is a need to move from the current dominant focus on the national level to the regional level when addressing energy system design and analysis in order to achieve sustainable energy systems (Ostergaard and Duic 2014). Furthermore, there are regional-based behaviours and practices that can significantly inform the national approach to renewable energy.

There are two ways of considering regional renewable energy practice and innovation: one focuses on the local community as a unit and another focuses on the local government and its role as a political and administrative unit. Hielscher et al. (2011) demonstrate that regionally-based and local community projects tend to be more effective in promoting renewable energy and behaviour changes compared to top-down, nationally initiated projects. While such projects have the potential to address social, cultural and economic barriers more effectively, their success depends on informed and engaged citizens and effective communication. As a result, local governments can often face challenges in supporting innovation, which can be addressed through the development of local technology policy (Rygg 2014).

The South West Region

Table 8.3 shows the progress that has been made against all forms of renewable energy by the top five UK regions during 2016. While Yorkshire and Humberside and the North Sea have the highest overall generation at just under 11,000 and 9682 GWh respectively, it is the south west region which is considered to be at the forefront of all forms of renewables in the UK, with the highest onshore installed capacity of any region.

The North Sea region only generates energy through offshore wind, and the leading position of Yorkshire and the Humber is heavily skewed due to the recent conversion in this region of the UK's biggest power station from coal to compressed wood pellets as part of Europe's largest decarbonisation project. This has reduced carbon dioxide emissions by

Table 8.3 Renewable energy progress in top 5 regions of the UK (GWh)

	Yorkshire and the Humber	North Sea	South West	South East	East of England
Onshore wind	1189		652	222	866
Offshore wind		9682			
Solar PV	381		2421	1625	1628
Energy from waste	1093		255	1195	87
Biomass	7912		987	920	1445
Other	325		1225	1242	1111
Total	10,900	9682	5540	5204	5137

Source Adapted by the author from magic.piktochart.com/output/16135495-renewable-energy-progress-report-2016

at least 80%. Currently, 65% of the electricity at the power station is generated from renewables (www.drax.com/technology/the-single-biggest-transformation-of-our-century/). However, Yorkshire and the Humber performs significantly less well against other forms of renewables in comparison to the south west region, particularly solar PV: 62% of the south west's renewable electricity is currently generated using solar PV. The south east is the second largest region in terms of land area but has the highest population of all UK regions. Its more urban nature perhaps reflects its greater use of Energy for Waste (EfW) compared to the south west where conurbations are typically smaller and more widely distributed.

The significant progress made by the south west region against multiple forms of renewable energy, as Table 8.3 demonstrates, can be largely attributed to the local support and governance provided by the South West Regional Development Agency (SWRDA). It was created with other RDAs in 1998 and since it was agreed that they should exist at arm's length from central government, each agency was able to create its own structures to achieve the following purposes:

- To further the economic regeneration of their area;
- To promote business efficiency, investment and competitiveness;
- To promote employment;
- To enhance the development and application of skills relevant to employment; and
- To contribute to the achievement of sustainable development in the UK (www.bristol.ac.uk/media-library/sites/red/migrated/documents/short-history.pdf).

While SWRDA's funding role was significant, as an impartial player, it was able to bring organisations together in neutral settings to enable collaborative ventures that could not have been formed otherwise. The agency was responsible for leading projects, such as the Bristol and Bath Science Park, the Cornwall Wave Hub, the East of Exeter Training Academy, the Centre for Additive Layer Manufacturing in Exeter and Swindon's hydrogen filling station, which all contributed to developing the South West economy. All of SWRDA projects can be considered visionary in nature. As economic development is a long-term process, the impact of these projects will take years to fully manifest itself (www.gov.uk/government/uploads/system/uploads/attachment_data/file/247051/0198.pdf).

Although regional economic strategies were heralded as a new policy instrument, they had to accommodate a wide range of inherited programmes and, despite their autonomy, were also required to reflect national strategies and priorities. The lack of joined up government policy on regional matters made balancing the expectations and aspirations of different departments and remaining faithful to regional and local needs and conditions, much more difficult (SWRDA 2011). While SWRDA, along with all other agencies, was abolished in 2012, its legacy is still strongly in evidence throughout the region. Initially the new Government stated that RDAs would be replaced by Local Enterprise Partnerships (LEPs). These are joint local authority and business partnerships which are intended to cover city regions or other 'natural economic areas' (SWRDA 2011). However, there have been issues with both implementation and management, and LEPs have attracted significant criticism due to their ineffectiveness.

Established in 2003 as part of SWRDA, Regensw is now an independent, not-for-profit organisation that upholds many of the principles and governance mechanisms developed by the original agency, believing that sustainable energy plays a vital role at the heart of a sustainable economy and thriving local communities. It fills the support gap left by SWRDA and uses its expertise to work with industry, communities and the public sector to change the way energy is generated, supplied and used. Through its extensive experience it supports entrepreneurial businesses to form partnerships, drives innovation, provides financing to companies and individuals to identify new opportunities, and actively engages and supports local people and communities.

Devon and Cornwall

Although the government has historically tended to focus on large-scale wind power, its support for other economically viable alternatives, such as biomass and solar, and for research into renewable types with future potential, such as wave and tidal, is also recognised. The natural resource required for each renewable type is invariably greatest in rural and/or remote regions of the UK (Hain et al. 2005), and the south west region of the UK has vast natural resources, complemented with expertise in specific areas, such as marine, which has led to more development in the areas of wave and tidal energy than in other UK regions (Hyman 2017).

Devon and Cornwall are the two leading counties for total installed renewable electricity and heat capacity in the region, achieving 780 and 755 Mega Watts (MW) respectively, with only Wiltshire coming close to these levels at 680 MW. They lead in renewables deployment and have the largest number of renewable energy projects in the region with Cornwall having 19,412 and Devon 34,494 projects in 2016 (Regensw 2016). The prevailing form of renewable energy is currently solar PV, followed by onshore wind power. In 2016, Devon saw almost double the number of new rooftop PV installations compared to any other county, and over 10% of homes in mid Devon have installed solar PV, the second highest level in England, while Cornwall has the highest percentage (30%) of its electricity generated by renewables (Regensw 2016).

Local and Community Initiatives

Community mobilisation is deemed essential to enable the transition to a decentralised energy supply, and the full exploitation of the potential of renewable energies (Wirth 2014). Local and community-based energy initiatives are considered important because they express a public sentiment that global environmental concerns deserve more serious political attention (Hisschemoller 2012). However, despite interest in community-generated energy displayed by industry, activists, policy makers and concerned individuals, there is currently limited academic literature addressing local renewable energy behaviours and practice. Academic research has given significant attention to the technology, and some to the social aspects of these forms of implementation, but largely in a fragmented way. As early as the 1970s, researchers argued that the transition to renewable energy was a matter of systematically addressing social and situational problems and that the 'soft issues' were key for the success of a renewable energy project (Ostergaard and Duic 2014). Almost forty years later, the debate that the energy transition is more than just a technological and/or an economic problem continues.

Local energy initiatives display a large variety in size, scope and organisation, which contributes to the complexity of this important perspective, and are associated with the small-scale level, such as a village, neighbourhood or town (Hisschemoller 2012). In recent years, local and community projects have flourished in the UK, building on a foundation of alternative energy initiatives launched in the 1970s, and benefiting from policy measures that support the transition to a low carbon economy. Community energy is a diverse field of activity that includes both energy generation and conservation projects (Seyfang et al. 2013). A key driver for community energy appears to be a lack of trust in energy companies and a desire for autonomy and self-control. Through energy cooperatives at the local scale consumers can become producers of their own energy, closing the gap between consumers, producers and shareholders (Stokman 2010).

There is a number of benefits associated with local renewable energy supply (Stephens Scown and Regensw 2016), which include:

- Enabling local energy producers to sell power directly to the local community without going through the wholesale market and, therefore, having more control over the price;
- Enabling local job creation and keeping the profits in the local economy;
- Ensuring greater control over energy bills and the ability to pass on savings to customers, helping to reduce fuel poverty;
- Helping communities meet their carbon and environmental objectives;
- Creating social enterprises that customers can trust;
- Building support for local renewable energy projects; and
- Overcoming barriers related to connection to the grid.

The role of local ownership in engendering public support, acceptance of or opposition to a renewables' project is a key issue in community energy, particularly with reference to wind energy projects. Studies indicate that the commercial installation of large-scale wind power plants makes local participation and ownership impossible, serving to boost public opposition, while in contrast, small-scale community-based projects receive strong levels of support from local people (Wirth 2014). In addition, local opposition to wind energy installations can be reduced through local engagement, participatory decision-making processes and the fair distribution of the resulting economic benefits (Agterbosch et al. 2009).

These approaches are reflected in the UK governmental toolkit for 'Delivering Community Benefits from Wind Energy Developments', which suggests three interconnected motives for wind energy developers: Being a Good Neighbour, which fits with commitments to corporate social responsibility (CSR) and being part of the community; Sharing the Rewards, so that local communities benefit from the rewards reaped from 'farming' the wind blowing across their locality; and Paying Compensation for the impact on the landscape and local amenity, and inconvenience caused by construction, where developers will provide appropriate funds from their budget either to the parties most affected, or to be disseminated by the local council. The latter is considered problematic as it implies liability for negative impacts, injuries or poor practice, and developers prefer to be seen to have good motives and positive community impacts (Cass et al. 2010).

At the beginning of 2015 there were more than 5000 community energy groups active in generating, managing, purchasing and consuming renewable energy across the UK (Stephens Scown and Regensw 2016). The Community Renewable Energy Network (CREN) is a key local energy initiative applicable in multiple regions of the UK. It refers to an electrical smart microgrid with mostly renewable electricity generation, owned and operated by a community for its supply of electricity and potential trading benefit (Tomc and Vassallo 2015). The south west has a high proportion of community projects compared to other regions of the UK and in relation to its smaller overall population (Seyfang et al. 2013) and according to Merlin Hyman (2017) the region has a long tradition of community activism. Devon and Cornwall again dominate in the number of community-based renewable energy projects they develop and implement.

A key example of successful community activism and implementation in the county of Cornwall is the Wadebridge Renewable Energy Network (WREN). In 1926, Wadebridge had its own Electricity Supply Company. Electricity production and supply was nationalised in the UK in the 1960s, and the resulting distribution through the National Grid meant Wadebridge lost its well-established energy independence. However, through recent community engagement WREN has implemented a plan to become energy self-sufficient once more. As articulated by Stephen Frankel of WREN 'The most profound and long-term impacts of community energy cannot be secured unless communities control their supply of energy' (Stephens Scown and Regensw 2016). As a result, WREN is building on its established community engagement and approaches to energy efficiency to secure the necessary income streams. It also controls supply arrangements, through the Wadebridge Energy Company, and aims to achieve renewable generation at the scale of consumption.

WREN has engaged with 10,000 people within the local community and worked collaboratively to establish a clear rationale and business case for developing their own energy network. They evaluated how much energy was being used within the town, how much it was potentially costing, based on 10,000 people each paying £1000 and established that as much as £10 million was going out of the community per annum. As a result of this assessment, community members realised that if they generated just a proportion of the total energy volume themselves the money could be kept locally and invested in the local economy. This prompted them to understand how they use energy, how they can produce it and what amount of energy could be generated from different forms of renewables, i.e. what would one wind turbine generate, what if they put solar on everyone's houses etc. As a result of acquiring this information, it was then possible to receive broad acceptance by the community (Hielscher et al. 2011).

A larger scale example of a community-based renewable energy project within the county of Devon is Plymouth Energy Community (PEC), a community benefit society founded in 2013. Its primary aim is to give people in the city the power to transform how they buy, generate and use power. As stated by the founder of PEC, Dave Garland, 'we set out to create a community of like-minded people who are committed to helping transform all things energy-related for the benefit of the local community and we are doing just that!'. The work of PEC focuses on three key energy goals (plymouthenergycommunity.com/about/aims-values):

- Reducing energy bills;
- Improving energy efficiency; and
- Generating a green energy supply.

In collaboration with a number of local organisations PEC has created tools and relationships to enable the community to achieve these goals, ensuring that they have the necessary power and knowledge to take action. They recognise the importance of putting the community at the heart of all decisions and measures, and that collective power can bring real change to Plymouth's energy future, which would stem from saving money, reducing the amount of energy used and generating their own energy from renewable sources (plymouthenergycommunity.com/about/aims-values).

As 13.4% of Plymouth households live in fuel poverty, Plymouth City Council, in line with its well-established co-operative ethos recognised that community energy could be a potential solution to this issue and it was responsible for the initial set up of PEC, through funding and the development of a business plan. In July 2013, the council handed the entire control of PEC over to a board of volunteer directors from across the community but continued to provide expertise from its low carbon business, finance, legal and HR teams. Alongside core services, such as advice and insulation schemes, PEC also funds and builds community-owned renewable energy projects in Plymouth. They include free solar panels for schools and community buildings and the installation of a solar array on derelict land. For the latter project PEC had to make sure they did it as quickly as possible, in 2015, before the government cut subsidies to expand the utilisation of renewables. It is worth noting that such national policy of changes to subsidies and tax relief indicate the challenges faced by community groups (plymouthenergycommunity.com/about/aims-values).

Barriers to Community Renewable Energy Projects

Community renewable energy projects receive overwhelming backing from the UK public (The Co-operative Energy 2016) and can achieve sustained levels of success. However, despite the recognised relevance of the local perspective for the transition to renewable energy and the clear success of initiatives, such as WREN and PEC, there are many barriers to regional community energy projects, to include laws and regulations, tax regimes, energy infrastructure and available expertise, and the behaviour of the 'green' financial sector, which can put local energy initiatives in a position of disadvantage (Hisschemoller 2012).

The community energy sector is exposed to a variety of barriers that are typical within the renewable energy industry but have a stronger impact on small local initiatives. These barriers include the changes to the subsidy regimes; construction risks; limited access to finance; regulatory barriers; and a requirement for niche expertise. Subsidy changes in particular have had a dramatic, negative impact on the viability and project implementation in the community energy sector. In 2016, the survey of the sector indicated that 44 out of 144 projects were considered stalled or currently inactive, and 48% of the community energy groups stated that FiT changes represented the main barrier to their project, with 34% also noting barriers to financing, often as a result of a lack of support in the form of a subsidy. Further barriers to project development included planning issues (25%), engineering issues (11%), lack of expertise and local opposition (7%) (www.communityenergyengland.org/files/document/51/1499247266_CommunityEnergy-StateoftheSectorReport.pdf).

This reflects the difficulty for community energy groups to establish the economic viability and technical availability required for their projects (Walker 2008). As well as ongoing cuts to FITs, there have also been broken promises around tax relief for renewable projects and in particular, the exclusion of cooperatives from the new Innovative Finance ISA (The Co-operative Energy 2016). The latter intended to offer the promise of a good return, sheltered from tax, to investors willing to take on the higher risks of the peer-to-peer (P2P) finance market (www.ft.com/content/2bf10faa-14cb-11e8-9c33-02f893d608c2). Due to this lack of funding, community groups face difficulties in raising sufficient capital, especially for the early high-risk costs at the pre-planning stage. Although government grants and social investment funds can offset some of the costs and risks, financial challenges remain, especially when government rules are inconsistent and promises are often uncertain (Bomberg and McEwen 2012). There is an ongoing debate on whether current RO policies are sufficiently effective, and it is felt that the UK national policy framework is inconsistent and only provides short-term incentives (Hisschemoller 2012) that are not strong enough to stimulate growth in community-led renewables' projects (Walker 2008).

From a legal perspective there are often specific constraints and conditions under which community organisations or projects can operate (Walker 2008). Some government capital funding programmes have focused on the not-for-profit status of community groups to ensure that public funding could be used without contravening EU rules (Walker et al. 2010). There are also the risks and difficulties associated with obtaining planning permission for solar and wind technology, in particular, due to resistance from the local authority and/or public opposition (Walker 2008). Together with the reduction in government subsidies these planning constraints have made it particularly difficult to establish wind co-operatives in recent years, although the importance of mobilising local supporters of wind energy is strongly recognised within the sector (The Co-operative Energy 2016).

Finally, there are also barriers to market entry and network connection that would enable communities to realise their income-generating potential. This is illustrated by a current lack of incentives for network providers to connect the community energy producers to the grid and the difficulty of smaller producers to access green energy certificates (Walker 2008). The difficulty of selling the electricity they generate on retail markets has been experienced by many community energy groups. There is a recognised need to make it as easy as possible for customers to buy into community energy supply arrangements (The Co-operative Energy 2016). As articulated by Merlin Hyman of Regensw (2017) 'when you're developing new things, new business models, breaking the mould then it's difficult; there are regulatory challenges, public acceptance challenges, the question of what is the route to market?' Energy storage is considered a key potential solution to this market problem for community renewables, as it can address the gap between potential generation and locational constraints (The Co-operative Energy 2016).

Conclusion and Future Research

Through the review of relevant research on the theme of renewable energy, together with a range of data from county councils, government, regional energy reports and independent sources, this chapter has illustrated the growing commitment of the UK to achieve a low carbon economy and significantly increase the share of energy from renewable sources. This is primarily due to the growing interest in ensuring energy security via increased energy generation from renewables. In addition, there are deep concerns regarding fossil fuel depletion and climate change (Wood and Dow 2011). However, the review has highlighted the limitations of addressing these issues by focusing on the national level, where energy policy decision-making has historically occurred. Furthermore, the chapter underlined the importance of understanding local and regional implementation and community engagement.

By focusing on the south west of the UK, a region that is recognised as a leader in renewable energy and resource efficiency, this chapter has emphasised the relevance of the regional perspective for enabling a successful transition to renewable energy at the national level (Ostergaard and Duic 2014). The south west demonstrates a deep commitment and ability to harness its abundant natural resources, supported by relevant skills and technology, and recognises the positive impact renewable energy can have on the local economy. However, the findings also indicate the key role of local governance, while much of the region's success and legacies are directly attributable to the SWRDA, which was abolished in 2012. The structured and coordinated support provided by this agency over a 14-year period encouraged and enabled collaboration between multiple stakeholders, through the effective dissemination of information and engendering public support for renewable energy projects (Wirth 2014).

The support gap left by the SWRDA in this region has subsequently been filled by Regensw, an independent, not-for-profit organisation. This further emphasises the need for clear principles, mechanisms and expertise for renewable energy, particularly at the community-level. The counties of Devon and Cornwall illustrate the key role played by community-based energy initiatives in transitioning to decentralised renewable energy provision across the UK (Seyfang et al. 2013), which was demonstrated by the case studies of WREN and PEC. Together with using its vast natural resources, the region is also famous for its traditionally high level of community activism, which better positions it, compared to other regions, to implement green energy practices at the local level. However, it still faces many recognised institutional and bureaucratic barriers, such as lack of financial support and incentives, legal and planning constraints, producers' limited access to the market and the grid, and lack of technical expertise (Hisschemoller 2012).

For other UK regions looking to implement, develop and increase their generation of renewable energy, there are a number of learning points. Despite the challenges it faces, the south west is highly proactive in addressing the often-negative view of community-only initiatives. This view is taken by the government and results in the lack of governance; therefore, the region is working to take control of its renewable energy generation. As this chapter has shown, Devon and Cornwall are particularly resilient and adaptable in their approaches and are often specifically motivated by the lack of government support for local projects. A recent action undertaken by Cornwall, which other counties could replicate, aims to secure a devolution agreement with central government to remove barriers to local energy generation and increase the associated local benefits. As 98% of Cornwall's spend on energy currently leaves the county, it has focused on initiatives to reverse this trend and address the recognised barriers. These include operating a revolving fund for community groups; developing a local policy framework that prioritises community energy; and providing guidance for Neighbourhood Planning groups looking to support their own community projects (The Co-operative Energy 2016).

There is also the need for regions and counties to fully and collaboratively engage with individuals, communities and organisations, and provide the opportunity for the right kinds of discussions and innovative solutions. As Merlin Hyman (2017) recognises 'that's one of the things the south west has got right, having a space where people can come together, where they can convene people, tackle barriers. It's one of the reasons why things have gone well here.' With the loss of governmental driven support through regional agencies, such as SWRDA, these spaces and opportunities can be created by working with independent organisations such as Regen, or by local authorities and councils developing their own renewable energy networks.

In the short-term this can provide a focus for action that is region- or community-specific and can harness funding, knowledge and expertise from multiple sources. As illustrated by the PEC case study where local authorities/councils have provided resources, time and dedicated, experienced staff to supporting community energy, there has been sustained development of community energy organisations and their projects. Local authorities can also assist with the financing of projects through direct loans to community energy organisations and can help to secure third-party funding, such as charitable grants and commercial loans (www.communityenergyengland.org/files/document/51/1499247266_CommunityEnergy-StateoftheSectorReport.pdf). The supporting roles played by local authorities can consequently help them to implement their wider renewable energy strategies and achieve certain objectives. In the long term, there is the need to establish well-supported and coherent strategies and business models for renewable energy generation at the local, regional and national level, and therefore, enable a structured and coordinated progress to the 70% level of renewable energy that the UK will require to meet its carbon targets (Hyman 2017).

This chapter has indicated an imperative for more research on the local perspective of renewable energy initiatives and implementation to understand the contribution made by communities and the benefits and challenges they encounter, and what local support mechanisms are required to ensure success. Future research requires the application of multiple theoretical lenses to develop a more holistic view of renewable energy behaviours and practices. These theoretical frameworks should permit to capture the key tangible aspects of technology, policy and governance, as well as the more intangible human and relational components that communities can harness to initiate and manage their own projects. Case studies will be appropriate for exploratory research as they enable quantitative and qualitative perspectives, and are a means of studying dynamic, emerging phenomena, practices and concepts, and offer creative insights and high validity with practitioners (Eisenhardt and Graebner 2007).

Acknowledgements The author acknowledges the contribution of Merlin Hyman to this work.

References

Agterbosch, S., Meertens, R. M., & Vermeulen, W. J. V. (2009). The relative importance of social and institutional conditions in the planning of wind power projects. *Renewable and Sustainable Energy Reviews, 13*(2), 393–405.

Bomberg, E., & McEwen, N. (2012). Mobilizing community energy. *Energy Policy, 51,* 435–444.

Cass, N., Walker, G., & Devine-Wright, P. (2010). Good neighbours, public relations and bribes: The politics and perceptions of community benefit provision in renewable energy development in the UK. *Journal of Environmental Policy & Planning, 12*(3), 255–275.

Eisenhardt, K. M., & Graebner, M. E. (2007). Theory building from cases: Opportunities and challenges. *Academy of Management Journal, 50,* 25–32.

Ellabban, O., Haitham, A.-R., & Blaabjerg, F. (2014). Renewable energy resources: Current status, future prospects and their enabling technology. *Renewable and Sustainable Energy Reviews, 39*(C), 748–764.

Hain, J. J., Ault, G. W., Galloway, S. J., Cruden, A., & McDonald, J. R. (2005). Additional renewable energy growth through small-scale community oriented energy policies. *Energy Policy, 33,* 1199–1212.

Hielscher, S., Seyfang, G., & Smith, A. (2011). *Community innovation for sustainable energy* (CSERGE Working Paper). CSERGE (Centre for Social and Economic Research on the Global Environment). http://www.cserge.ac.uk/sites/default/files/2011-03.pdf.

Hisschemoller, M. (2012). Local energy initiatives cannot make a difference, unless. *Journal of Integrative Environmental Sciences, 9*(3), 123–129.

Hyman, M. (2017). Interviewed by Dr Alison Ashby, 1st August.
Kitzing, L., & Weber, C. (2015). Support mechanisms for renewables: How risk exposure influences investment incentives. *International Journal of Sustainable Energy Planning and Management, 7*(1), 117–134.
Ostergaard, P. A., & Duic, N. (2014). Sustainable energy, water and environmental systems. *International Journal of Sustainable Energy Planning and Management, 3,* 1–4.
Panwar, S. C., & Kaushik, S. K. (2011). Role of renewable energy sources in environmental protection: A review. *Renewable and Sustainable Energy Reviews, 15,* 1513–1524.
Regensw. (2016). *Renewable energy: A local progress report of England.* https://www.regensw.co.uk/Handlers/Download.ashx?IDMF=19345491-c8de-4cb2-99b3-4273f521ee38.
Rygg, B. J. (2014). Paving the way for heat. Local government policies for developing bioenergy in Norway. *International Journal of Sustainable Energy Planning and Management, 4,* 57–70.
Seyfang, G., Park, J. J., & Smith, A. (2013). A thousand flowers blooming? An examination of community energy in the UK. *Energy Policy, 61,* 977–989.
Stephens Scown, & Regensw. (2016, March). *Local supply: Options for selling your energy locally.* https://www.regensw.co.uk/Handlers/Download.ashx?IDMF=9b4bd983-7ee6-4b65-b45f-25d22c5f277d.
Stokman, F. N. (2010). Microfoundation of and fundamental solutions for sustainability connecting human behaviour and Gaia theory (unpublished manuscript).
SWRDA. (2011). *Reflections and lessons.* http://www.bristol.ac.uk/media-library/sites/red/migrated/documents/reflections-and-lessons.pdf.
The Co-operative Energy. (2016). Community Energy Conference 2016—Post event report.
Tomc, E., & Vassallo, A. M. (2015). Community renewable energy networks in urban contexts: The need for a holistic approach. *International Journal of Sustainable Energy Planning and Management, 8,* 31–42.
Walker, G. (2008). What are the barriers and incentives for community-owned means of energy production and use? *Energy Policy, 36,* 4401–4405.
Walker, G., Devine-Wright, P., Hunter, S., High, H., & Evans, B. (2010). Trust and community: Exploring the meanings, contexts and dynamics of community renewable energy. *Energy Policy, 38,* 2655–2663.
Wirth, S. (2014). Communities matter: Institutional preconditions for community renewable energy. *Energy Policy, 70,* 236–246.

Wood, G., & Dow, S. (2011). What lessons have been learned in reforming the Renewables Obligation? An analysis of internal and external failures in UK renewable energy policy. *Energy Policy, 39*, 2228–2244.
www.bristol.ac.uk/media-library/sites/red/migrated/documents/short-history.pdf.
www.communityenergyengland.org/files/document/51/1499247266_CommunityEnergy-StateoftheSectorReport.pdf.
www.drax.com/technology/the-single-biggest-transformation-of-our-century/.
www.gov.uk/government/uploads/system/uploads/attachment_data/file/47871/25-nat-ren-energy-action-plan.pdf.
www.gov.uk/government/uploads/system/uploads/attachment_data/file/622815/Renewable_energy_in_2016.pdf.
www.gov.uk/government/uploads/system/uploads/attachment_data/file/556271/Renewable_electricity.pdf.
www.gov.uk/government/uploads/system/uploads/attachment_data/file/247051/0198.pdf.
www.gov.uk/government/statistics/regional-renewable-statistics.
www.plymouthenergycommunity.com/about/aims-values.

9

Local Government Programmes for Energy-Efficient Solutions: Procurement Options and Governance Challenges in Street-Lighting Renovation

Mihaela Grubišić Šeba

Higher energy efficiency is one of the key elements of the European Union's 2020 strategy for smart, sustainable and inclusive growth. For this reason, many EU member countries and the prospective EU accession countries are applying energy efficiency programmes on their own and with financial injections from international donors.

Street-lighting renovation is probably one of the easiest ways to achieve higher energy efficiency in a very short time. This is because most street lamps and installations are outdated and, therefore, energy inefficient. Street lighting accounts for 30–50% of the entire power consumption of public authorities, whereby maintenance and operation costs (which include electricity cost) during the useful economic life of street lighting climb to 85% of its total costs.

This chapter aims to analyse how different financing approaches to energy efficiency projects contribute to public policy on street-lighting renovation. There is typically a top-down approach in policy design in

M. G. Šeba (✉)
BizVision d.o.o. for business advisory, Zagreb, Croatia
e-mail: mihaela.grubisic-seba@bizvision.hr

N. Mouraviev and A. Koulouri (eds.), *Energy Security*,
https://doi.org/10.1007/978-3-030-01033-1_9

the street-lighting renovation field, which starts from the target savings that need to be achieved at the local level. The accompanying financing models for street lighting include traditional public procurement, energy performance contracts (EPCs or EnPCs) implemented by energy service companies (ESCOs), and public–private partnership models. Each model can be financed by own funds (i.e. equity), loans, grants or any combination of financial instruments. Each model has a different decision-making procedure for public authorities responsible for proper street lighting functioning and street-lighting renovation. In some nations, electricity operators or companies that manage various parts of public infrastructure (motorways, railways) are in charge of public lighting within the boundaries of the infrastructural facility under their management. While the operators of public infrastructure can be either publicly or privately owned, the ultimate responsibility for public lighting rests with the public authorities. In the countries with a significant burden of public debt, the local authorities have to take into account their own and the national current fiscal position as well as consider the outlook for future fiscal conditions. It is often the case that the latter determines the optimal financing model of street-lighting renovation. For indebted nations, private initiatives are highly welcome, either through the energy service or public–private partnership contracting.

When pursuing street-lighting renovation, the local authorities need to go through the public procurement process. The required project documentation is very detailed, as economically, the most favourable bid needs to be accepted. Yet, the financial and administrative capacities of local authorities vary, and they are typically insufficient for running the street-lighting renovation project smoothly. Public procurement is supposed to help, but sometimes it complicates the timely implementation of small value projects. The existing contracts for street-lighting maintenance, which are often entrusted to local vendors, also experience similar difficulties in the implementation process.

The criteria for calculating the economically most favourable bid vary and local authorities are free to set them on their own. This chapter analyses what criteria are used most often and why. It also investigates what model(s) of street-lighting renovation are preferred by local authorities and which models the private sector tends to choose.

The Significance of Street Lighting in Local Governance and Local Finance

Public lighting purchase, installation, functioning, maintenance and decommissioning is the ultimate responsibility of local authorities in most countries prescribed by-laws on communal services, and it uses as much as 60% of a municipalities' overall budget (European Commission 2013). Their responsibility can be considered from the perspective of the legal owner of the street lighting, on the one hand, and from the perspective of the ultimate payer of the bills related to street lighting, on the other. Technically, the purchase of street lamps and other street-lighting installations are conducted by or on behalf of the government, which amounts to about 16% of the European GDP in total (Traverso et al. 2017).

A public street light is defined as a fixed lighting installation, to provide good visibility to users of outdoor public traffic areas during the hours of darkness to support traffic safety, traffic flow and public security (Van Tichelen et al. 2007). It includes functional lighting of pedestrian and cycle paths as well as roadway lighting. The first street illumination is related to the gaslights of London's Pall Mall and dates back to 1807 (Elberg and Woods 2017). More than 200 years later, the European Commission (2013) estimates that there are 90 million traditional street lights on its territory alone, of which 75% were older than 25 years. The EU-28 has about 5.5 million km of roads, of which 43% are lit. Among them are 8500 km of motorways, 34,000 km of national or main roads, 272,000 km of regional roads, and 1,350,000 km of other roads (Traverso et al. 2017). Street lighting typically accounts for 30–50% of total electricity consumption of municipalities, 15% of global electricity consumption, and 6% of global greenhouse gas emissions (UNEP/GEF 2013). The energy consumption for street lighting was estimated at 35 TWh in EU-25 in 2005, which was about 1.3% of the total energy consumed annually (Van Tichelen et al. 2007). An ambitious target of the EU is to reduce GHG emissions by at least 20% compared to the 1990 level, increase the share of renewable energy sources by 20% and achieve a 20% increase in energy efficiency by 2020 (EC 2010). Street-lighting installations are classified as simple constructions, meaning that the implementation of these projects is shorter and require less financial resources compared to more complex energy efficiency projects in public buildings or manufacturing facilities. In addition to ensuring the energy (and public cost) savings, the goals of street-lighting renovation include better light quality and visibility, less light pollution and improved street safety and security.

Street-lighting installations include street-lighting systems (composed of control gear, lamps, luminaries and an electric power system) and street-lighting poles. Street-lighting renovation can pursue the following energy efficiency tasks:

- Modernisation (and relocation) of public lighting switchgear assemblies;
- Modernisation of the distribution network, with the aim of increasing safety and reducing network losses;
- Relocation (and installation) of the lighting poles, with the aim of optimal positioning of the luminaires;
- Installation of new luminaires, with fulfilment of all requirements in the field of energy efficiency, protection against light pollution, illumination and other lighting standards; and
- Replacement of the light sources, with fulfilment of all energy efficiency requirements, protection against light pollution, illumination and other lighting standards.

Of all these energy efficiency measures, street lamps are most responsible for energy consumption and/or savings. The technology of street lamps has rapidly advanced from manually operated mercury-filled street lighting, to various types of light-emitting diode (LED) street lamps, remote controlled illumination and other smart city solutions. Switching the obsolete street lamps into LED is not only the question of lower GHG (CO_2) emissions, but the question of reduction in light pollution, which is especially evident in large cities. LED street lamps cut energy costs by 50–80% and can last up to 25 times longer than traditional incandescent lights. A comparison of the technical features can be obtained from various manufacturers as well as from independent sources (http://www.ecosolenergy.com/documents/Why_Go_for_LED.pdf; https://www.noao.edu/education/QLTkit/ACTIVITY_Documents/Energy/TypesofLights.pdf).

The environmental impact of different street-lighting technologies is well described by Welz et al. (2011). About a 50% reduction in energy costs stems from the replacement of the conventional street lamps with LED, and a further 30% of savings is attributed to smart city solutions that can enhance public safety, traffic management, health and comfort. LED technical solutions may concern modern street-lighting management (as shown in the left-hand column of Table 9.1), which can, typically in larger cities, be accompanied by the smart city solutions (these are shown in the right-hand column of Table 9.1).

Even though it is still perceived as very expensive, the initial cost of LED street lamps was reduced by as much as 85% between 2008 and 2013 (Matulka and Wood 2013). Basic remote controls of street lighting, such as remote on-off control, dimming and scheduling, are now standard features of the new lamps offered in the market, while additional possibilities require low to medium upfront or ongoing costs. The most expensive smart city solutions are those that require broadband, especially video transmission. However, the costs of street-lighting management and smart city solutions need to be compared, not only with available technological solutions, but also with the costs of a number of public procurement procedures for multiple smart city solutions, as well as with the estimated salaries and the number of the employees paid from the public budget after these solutions have been installed. In other words, the total cost of ownership or life cycle costing calculation should be applied when comparing net benefits of modern

Table 9.1 Street-lighting management and smart city solutions

Street-lighting management	Smart city solutions
• Remote on-off control, dimming and scheduling function • Real time energy consumption monitoring and billing • Remote performance monitoring • Colour controls for different purposes • Adaptive lighting according to the traffic density and/or street activity • Emergency response function	• Environmental/air quality monitoring • Traffic monitoring • Smart urban monitoring (parking, waste management, gunshot detection) • Broadband (3G/4G cellular, public WiFi, point-to-multipoint connectivity, traffic light controls, HD CCTV)

Source Adapted by the author from Elberg and Woods (2017)

street-lighting installations. Depending on the technology used, the capital costs of street-lighting installations can be very high, particularly if not only lamps, but poles and luminaires are installed or changed. Due to ongoing electricity costs, operational costs have been the major part of the total cost of ownership of street lighting. In addition to electricity costs, environmental impact assessments need to be completed. Between 85 and 90% of the environmental impact of the street-lighting lamps is ascribed to its use, and it also applies to LED lamps. This is due to the high percentage of fossil fuels in electricity generation (Van Tichelen 2007; Hartley et al. 2009; IEA 2014). The two most significant parameters contributing to the environmental impact are luminous efficacy (lm/W) and useful life (hours of operation during lifetime). Luminous efficacy measures the quality of visible light produced by the light source. LED lights typically last longer than conventional street lights, and recently they reached the level of up to 100,000 hours of operation. The lower the lamp's energy consumption and the higher its luminous efficacy and lifetime, the lower is its environmental impact.

For a local public authority, the most important reasons for street-lighting renovation are its obsolescence, demand of the citizens for greater traffic and pedestrian safety, rule of law and high public spending for energy. From the national perspective, the levels of public spending in general and energy spending in particular, play a critical role in setting the national goals for a decrease in energy consumption. The responsibility for proper street lighting functioning lies ultimately with the local authorities, as described by communal services laws. Yet the responsibility for overall energy consumption, environmental protection, light pollution, and public spending rests with the central government that is charged with setting the energy efficiency goals.

Financial Models for Street-Lighting Renovation

Street-lighting renovation is always paid for from the public budget, although the financial models vary. For the local authorities that have sufficient funds, there is an option to make a one-off payment after the renovation to the public service contractor. In case of borrowing the funds from a lending institution, the payment can also be done as a lump sum immediately after the renovation has been completed. Both options represent lighting contracting via traditional public procurement models. Light supply contracting places the entire care for street lighting functioning and installation in the hands of a private partner (contractor). The latter is responsible for construction, financing, operation and maintenance of street lighting and even for electricity purchase.

In between the two options there is a typical contemporary approach in street-lighting renovation, which is contracting an energy service with a private partner for a certain period of time. The private partner is then responsible for financing and construction on the certain area and it guarantees energy savings to the local authority during the energy service period, in return for a periodical, typically monthly, fee. The operation typically rests with the local authorities, while the private partner is responsible for the maintenance of the new lamps within the guarantee period. This is commonly known as energy performance contracting (EPC).

The Energy Efficiency Directive (2012, p. 11) defines EPC as; a contractual arrangement between the beneficiary (i.e. public authority) and the provider of an energy efficiency improvement measure (private partner/ESCO), verified and monitored during the whole term of the contract, where investments, i.e. work, supply or service, in that measure are paid for in relation to a contractually agreed level of energy efficiency improvement or other agreed energy performance criterion, such as financial savings'. The Energy Efficiency Directive (2012, p. 11) further defines an ESCO as 'a natural or legal person who delivers energy services or other energy efficiency improvement measures in a final customer's facility or premises'.

In EPC, a private partner guarantees the energy savings to the public authority. It finances the investment in the street-lighting installations upfront and recovers its investment over the contract term. To avoid the influence of price fluctuations on the fee size during the guaranteed savings' period, energy savings are calculated against the reference consumption of the electricity of the public authority, measured before entering into the EPC contract. The local authority operates the system of street lighting during the contract duration and savings are calculated periodically. The EPC model with guaranteed energy savings goes hand in hand with the third-party financing, in which ESCO bears investment and credit risk, the local authority assumes operational and user behaviour risk, while other risks, such as inflation or foreign exchange risk, are typically shared between the ESCO and the local authority (Bertoldi et al. 2014). The contract term is divided into a project implementation phase and a guaranteed savings phase. In the first phase, the project is implemented and there is no payment from the local authority to the ESCO. In the second phase of the contract, a stream of payments flows from local authorities to the ESCO if the guaranteed energy savings are achieved. For instance, if the guaranteed 30% of energy savings have materialised in a previous month, then the private partner has a right to claim its fee in the current month. Sometimes, a shared savings mechanism is agreed, meaning that energy savings above the minimum guaranteed threshold are attributed to both contractual parties by applying a certain ratio. As far as the maintenance is concerned, the local authority can keep the existing contracts for street lighting with the local maintenance companies or in-house or give the entire maintenance of the street-lighting system to a private partner. While in EPC the investment in energy efficiency renovation is repaid from the savings in energy consumption (and sometimes maintenance) over the contracted period, in light supply, contracting savings might not be sufficient to cover the overall investment and operation costs. In other words, in EPC the fee the local authority pays to the private partner remain the same as if there was no street-lighting renovation for the period of contract, whereby one part of the fee covers the electricity costs and the other part of the fee makes up for the capital costs of street-lighting renovation.

A guaranteed savings' energy performance contract in street-lighting resembles a financial leasing contract in terms of accounting treatment of the street-lighting assets. Assets are kept and depreciated in the books of the ESCO until the end of the EPC, when they are transferred to the local authority that remains its owner all the time. Legal ownership rests with the local authority, while economic ownership belongs to the ESCO that depreciates street-lighting assets in its books over the contract term. Hereby the economic ownership means that the street lights are kept in the books of the contractor until the end of the contract, although local authorities remain the ultimate legal owners of the street-lighting infrastructure.

The similarities and differences between the typical street renovation models are shown in Table 9.2 with respect to certain project phases and guaranteed energy savings and economic ownership of the project features. The complexity of the public procurement procedure increases from the left to the right-hand part of the table. The lighting contracting is the simplest, EPC is of moderate complexity, while light supply contracting is the most complex. Light supply contracting refers to the transfer of overall local public authority's responsibility for street lighting functioning to the private partner. However, payment of the fee, i.e. financial risk, rests with the public authority. Even though the private partner guarantees energy savings in light

Table 9.2 Public authority vs. private partner's responsibility in different contractual types of street-lighting renovation

Project phase/feature	Street-lighting renovation responsibility		
	Lighting contracting	EPC	PPP/Light supply contracting
Financing	Public authority	Private partner	Private partner
Design	Private partner	Private partner	Private partner
Installation	Private partner	Private partner	Private partner
Operation	Public authority	Private partner and/or public authority	Private partner
Electricity purchase	Public authority	Public authority	Private partner
Maintenance	Public authority	Private partner and/or public authority	Private partner
Guaranteed savings in electricity consumption	None	Private partner	Private partner
Economic ownership	Public partner	Private partner	Private partner

Source Compiled by the author

supply contracting, there is no guarantee of the price of electricity as it is bought at the discretion of the private partner. The EPC and light supply contracting are often referred to as public–private partnerships. However, the light supply contracting is much closer to a public–private partnership as it defines the quality of service, rather than parameters of the energy service delivery. However, the difference is sometimes in the detail and it often stems from how a public–private partnership is defined by legislation in a certain country. A more detailed comparison between EPC and energy supply contracts can be found in Boza-Kiss et al. (2017).

While lighting contracting is the concept the public authorities are most familiar with, EPC gained its popularity due to offering guaranteed savings in electricity consumption and the financing of street-lighting renovation by the private partner, which is preferred by indebted public authorities. However, the contract execution is as good as the contract's clauses. If the contract clearly specifies the rights and obligations of the parties, then its execution goes smoothly, even if problems during implementation occur, or the equipment turns out to be of poorer quality than agreed, or if the savings are smaller than expected. Most governments and international financial institutions take into account the limited know-how of public authorities in street lighting, offering publicly the technical guides for street-lighting renovation contracting or direct technical assistance in tendering process for renovation projects. The examples of such technical assistance include ESCO project support provided by the EBRD (available at: http://wb-reep.org/eng/), or technical assistance for project preparation provided by the European Investment Bank's and European Commission's ELENA facility (available at: http://www.eib.org/products/advising/elena/index.htm).

The Relation Between Financial Models and Public Procurement Procedure

Each project has to be prepared before making a decision on the renovation of public lighting infrastructure. A prepared project includes an energy audit with a precise description of locations of street lights, technology, energy consumption and electricity costs, lifespan, and maintenance costs.

Figures 9.1, 9.2 and 9.3 graphically represent the contract and money flow in street-lighting renovation contracting, i.e. EPC and light supply contracting or a PPP. In all three figures, either the European Commission or EU institutions or governing ministry for energy efficiency (EE) goals can provide technical assistance to public authorities for project preparation. This technical assistance may be with or without a charge. Technical assistance is necessary as staff in many local authorities do not have the required technical, legal or financial expertise to deal with the preparation of even simpler projects, such as public street lighting. In addition, there is typically no one-size-fits-all model of the street-lighting renovation prescribed by the national government. The risks of allowing the local authorities to engage with the public projects on their own include: no check of the price and quantity of necessary luminaries; no check of the technical features of the lighting infrastructure; no analysis of the costs and whether they are necessary and the offer of a questionable price in public contracting, among others. If engaged in energy efficiency projects, lending institutions may require technical assistance as well as the evaluation of technical characteristics of energy-efficient projects themselves. However, such technical assistance is typically paid from the loan fee charged to the borrower of funds.

As Fig. 9.1 shows, a private partner performs only works, i.e. street-lighting renovation without any guarantee for its energy-efficient operation thereafter. Local authorities are obliged by the law to take care of functioning and financing the street lighting and they do so in practice. A private partner can finance the street-lighting renovation either with its own funds or with borrowed funds. However, as the design and installation periods are rather short, it may only engage with a bridge loan (i.e. a loan that provides liquidity in the short term). Not only is this the simplest, but also the cheapest model for local authorities if they buy technologically adequate street-lighting solution that will enable them to achieve required energy savings. In this case, the local authority will gradually recover the initial capital cost of street-lighting renovation with lower electricity and maintenance bills over the economic life of the street lamps.

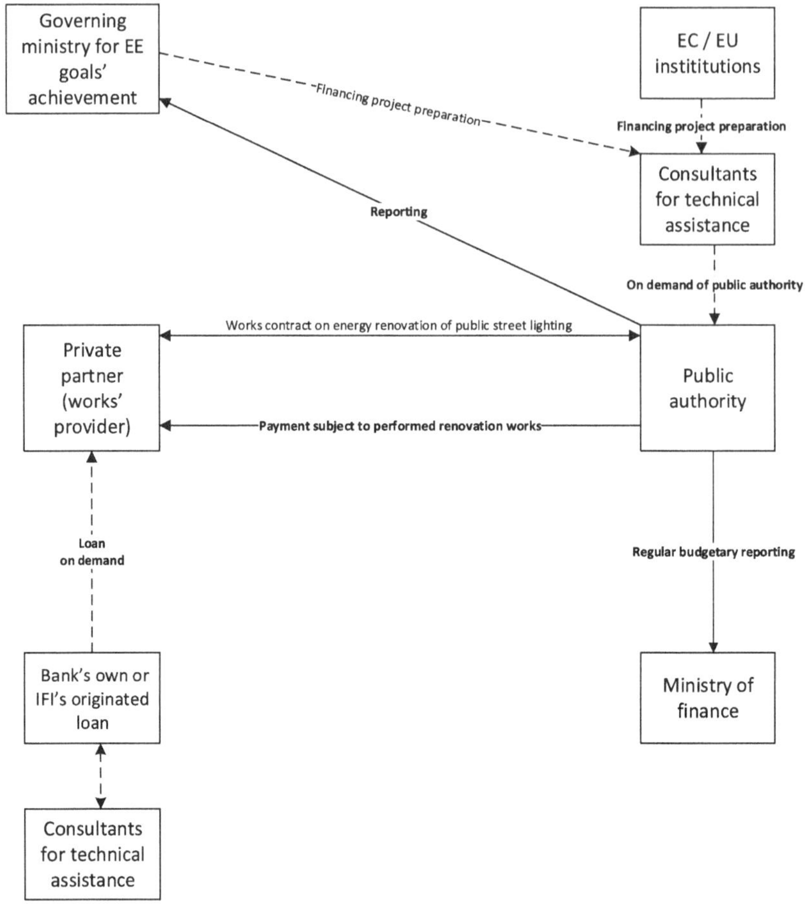

Fig. 9.1 Lighting contracting (traditional public procurement) (*Source* Compiled by the author)

The EPC concept illustrated in Fig. 9.2 differentiates from the lighting contracting as no fee is charged to the public authority until the street lighting starts functioning. The fee is subject to achieving the guaranteed savings. To protect themselves against the malperformance risk, private partners typically guarantee a lower percentage of possible energy savings, creating a safe margin for their earnings. The model proved to be efficient for replacement of street lamps and luminaries.

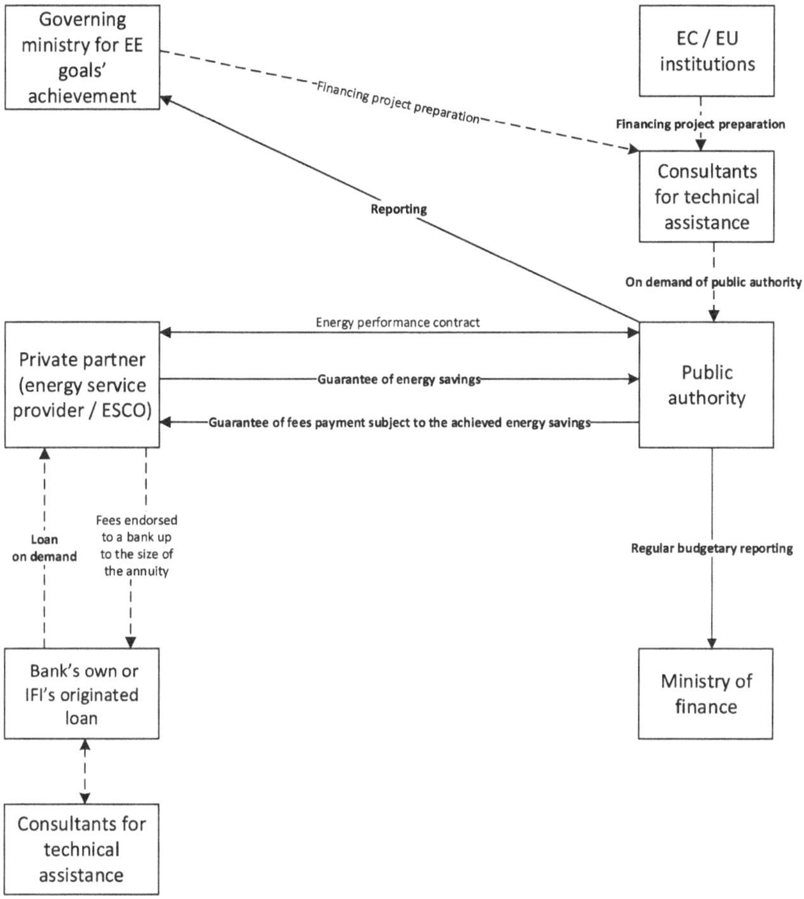

Fig. 9.2 Energy performance contracting (*Source* Compiled by the author)

Yet, higher capital costs of other street-lighting infrastructure prevent the model from being adequate in case of large-scale renovation. This is because private partners would not engage into the contract that requires them to guarantee energy savings and agree to receive the same fee (regardless of indexing due to inflation) for the period longer than eight to ten years.

Figure 9.3 shows a typical public–private partnership model in which a special purpose vehicle (SPV) is established for the purpose

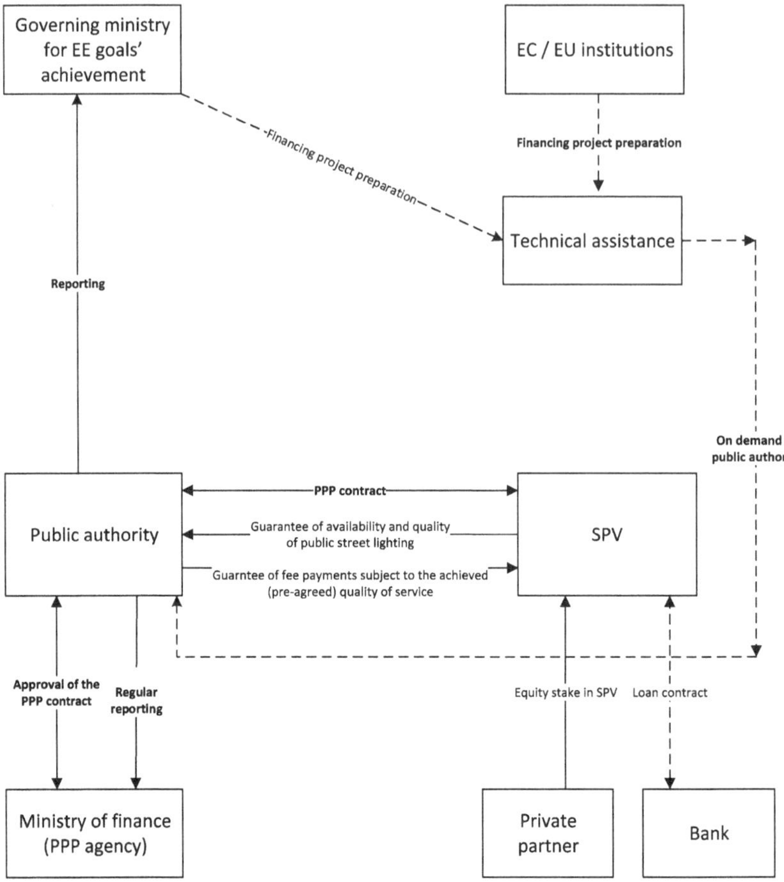

Fig. 9.3 PPP and light supply contracting (*Source* Compiled by the author)

of implementing the street-lighting renovation project. The private partner guarantees a certain quality of service, which may or may not include energy supply. The public partner pays for the availability of the street-lighting infrastructure. It is crucial that the contract is well defined and that it clearly sets out the rights and obligations of both contractual parties. This model is typically suitable for larger and longer infrastructural projects. Otherwise, it would not be feasible to engage into a lengthy public procurement procedure that requires selection of

Table 9.3 A comparison of three models of street-lighting renovation

Type of the renovation model	Contract duration	Responsibility for financing renovation	Responsibility for works	Responsibility for savings	Responsibility for operation
Traditional public procurement	Until works on lighting renovation are completed	Public partner	Private partner	Public partner	Public partner
EPC	Several years, typically up to 8 years	Private partner	Private partner	Private partner	Public partner
PPP	Several years (determined by the contract)	Private partner	Private partner	Private partner	Private partner

Source Compiled by the author

a private partner via competitive bidding and establishment of a special purpose vehicle for project implementation in line with the regulation on contracting public–private partnerships.

Table 9.3 provides a comparison of three typical models of street-lighting renovation. The level of involvement of the private partner is the highest in the PPP model, while in the traditional procurement model it is limited to the completion of public works. In all these models, the public authority is responsible for making payments. However, in traditional public procurement the public authority pays the fee immediately after the work has been completed, while all responsibility for street lighting functioning is then borne by the public partner. In the latter two models, the private partner guarantees savings (in EPC) or service quality (in a PPP), while the public authority accepts an obligation to pay the private partner in instalments either for the savings (in EPC) or for the service quality (in a PPP) achieved during the contract term.

Public Procurement Rules

The size of street-lighting renovation projects typically exceeds the minimum threshold for small value contracts that are exempt from the public procurement rules. Hence, a public procurement procedure needs to be followed. It means that an energy audit of street lighting needs to be attached in line with the other documentation, such as technical requirements, the contract's content and a cost list.

The quality of street lamps varies depending on the types of the lamps, manufacturers, technical features and environmental impact. Varying lamp quality is directly correlated with changing energy savings costs. Local governments face significant challenges on how to incorporate varying and at times, very high volumes of energy consumption that happen during certain holidays and special events (Christmas celebration, regional festivities, concerts) that can show anomalies in energy spending of the street lighting. When the consumption is greater, the savings are smaller, and, as the investment repayment term is bound to the savings achieved, smaller savings can endanger the expected payoff in the project. For this reason, the calculations of energy savings of new street lamps are compared to the so-called reference energy consumption. The latter is determined in the energy audit of street-lighting infrastructure and it is usually based on the average energy consumption of the local authority in the last three years prior to entering into street-lighting renovation contracts. Reference energy consumption is then compared with the projected energy consumption that results from the new street-lighting renovation project to determine the possible energy savings of the new, compared to the old, street-lighting infrastructure.

The mandatory public procurement rules throughout the EU-28 require that the economically most favourable bid must be selected, whereby the selection criteria are divided between the quantitative (price) and qualitative criteria (that include hours of operation, certain technical features that improve lighting quality and/or lower costs below the set level). Such criteria are typically set above the minimal technical requirements that the bidders need to fulfil if they want their bids to be further evaluated against quantitative and qualitative criteria. Even though the price criteria commonly take the greater part of the selection process of the bidders, the qualitative criteria are likely to adjust the decisions of the public authorities that are prone to select the lowest price bid. These qualitative criteria contribute to the energy efficiency policy goals at the local, regional, national and the EU levels. Apart from the standard regulation on public procurement that is the same for all public authorities, regardless of the sector, the rules for construction and renovation of public lighting systems are usually defined in energy efficiency regulation, environmental protection laws and laws against light pollution. The EU has also mandatory eco-design regulation, energy labelling regulation, regulation on the restriction of hazardous substances in electrical and electronic equipment and a regulation on recycling and waste disposal of electrical and electronic equipment. To help local authorities contract cost-effective street-lighting renovation, LED street lighting procurement and design guidelines have been developed by the Austrian energy agency within the EU funded project entitled Premium Light Pro (Austrian Energy Agency 2017) and they are presented in Table 9.4. They complement voluntary green public procurement rules developed for street lighting and traffic signals (EC—DG-Environment 2012; Traverso et al. 2017). These criteria have been aligned with the European Standard EN 13201 Road Lighting (http://www.eib.org/products/advising/elena/index.htm), and they exclude tunnels, parking lots and sports installations which have their own performance requirements defined in other standards. EN 13201 distinguishes between main road lighting classes for certain types of roads—motorways, main or national roads, regional roads and roads for pedestrians and cyclists. Similar rules have been embedded in national legislations.

Further information on LED technology can be found in the Development Finance International's Manual (2014). The examples of public procurement criteria in the EU countries can be found in Valentová et al. (2012) and the Green Partnerships Approach to Street Lighting (2014). For street-lighting luminaires the CE mark of conformity issued by the manufacturers covers the conformity with the so-called Low Voltage Directive (2014/35/EU) and the Directive 2014/30/EU on the harmonisation of the laws of the member states relating to electromagnetic compatibility and some other directives. CE mark of European Communities signifies the manufacturers or responsible vendor self-declaration of conformity with the applicable regulation for energy-efficient and safety-oriented street lighting. A product with ENEC mark from another European country is treated as if it had been certified by the national inspection body in its own country. ENEC+ mark is available both for LEDs and traditional light sources. ENEC+ enables independent third-party certification of the product that is reassessed after three years. All certified products are listed in a publicly available database, which guarantees safety and performance quality for the lifetime of the certification, and it is accepted throughout the EU and beyond. Apart from the Energy Efficiency Directive, there is also a regulation on eco-design requirements for lamps and related equipment (Directive No. 1194/2012). These directives all take care of the efficient use of the public funds in street-lighting infrastructure, emphasising (a) energy efficiency, and (b) value for money and the quality of street-lighting infrastructure. When put together, the technical rules seem too complicated and sometimes redundant, yet they prevent

Table 9.4 Typical qualitative criteria for public procurement of LED street lighting

Characteristics	Description of the criteria
Luminous flux	The total amount of radiation emitted by a given light source that is visible for the human eye. It is measured by lumens (lm)
Luminous intensity	The spatial distribution of light measured as the luminous flux within a given solid angle from the light source. It is measured by candela cd, 1 cd = 1 lm/square radian
Illuminance	The total amount of light reaching a particular illuminated surface area. It is measured in lux, 1 lux = 1 lm/m^2
Luminance	The brightness of lit surfaces or objects as perceived by the human eye. It is measured in cd/m^2. Minimum luminance requirements are specified for road classes covering medium to high speed motorways
Glare	An unpleasant visual effect caused by unfavourable distribution of luminosity or high contrasts, forcing the eye to adjust rapidly. Disability glare is caused by the scattering of light in the eye which reduces contrast sensitivity, while discomfort glare triggers a subjective sensation of discomfort
Light colour	White light (e.g. 4000 K) may typically be preferred for complex road situations with different types of road users involved (e.g. cars, cyclists, pedestrians). In contrast, lower, warmer colour temperatures may be preferred for domestic areas. Colour preference varies across the countries. Blue-emitting LEDs currently have the highest efficiency of all LED types, with a power conversion ratio of 55%. The remaining 45% is transformed into heat
Colour rendering	Is represented by the colour rendering index (CRI). Maximum index value is 100 while lighting systems with colour rendering of 80 or better are suitable for good facial recognition
Colour maintenance	Ageing LED modules may change their colour temperature and colour coordinates and very few manufacturers offer warranties for colour maintenance over the lifetime of LED
Light pollution	One way of reducing light pollution is to use luminaires which direct the light only on the areas to be illuminated

(continued)

Table 9.4 (continued)

Characteristics	Description of the criteria
Ingress protection	The resistance of luminaires against foreign matter is indicated by the so-called Ingress Protection (IP) code. For street lighting, IP65 luminaires should be used to ensure sufficient resistance to dust, particulates and in-clement weather
Mechanical impact	The resistance of luminaires to mechanical impacts is indicated by their mechanical Impact (IK) code, a number defined by the IEC 62262 standard. A minimum of IK08 is recommended
Voltage protection	Resistance to transient over-voltages (increases in voltage above the standard design voltage that last a few micro- or milliseconds). Many street-lighting projects mandate overvoltage protection up to 10 kV, although EN 61547 specifies only a 0.5 kV phase to neutral wire/earth
Efficiency	The total efficiency of LED lighting systems not only depends on the LED module efficacy, but also on the luminaire, the light control system and the overall lighting system design. Typical energy efficiency for LED is 150 lm/W
Lifetime	The average rated life Lx specifies the time it takes until the average LED module provides less than x per cent of its initial lumen output. For instance, L80 50,000 hours means that the lumen output of the module decreases by 20% after 50,000 hours of operation. The lifetime of the luminaire control gear also needs to be taken into account which is usually expressed as a percentage chance of failing within a particular time period, such as a failure rate of 0.1% per 1000 hours. LEDs generally have a lifetime of 100,000 hours or above
Energy performance indicators	EN 13201-5 describes the two energy performance metrics power density indicator (PDI) DP (measured in W/(l $\times$ m^2)) and the annual energy consumption indicator (AECI) DE (measured in (Wh)/m^2)

Source Adapted by the author from the Austrian Energy Agency (2017)

public authorities from contracting the projects that do not benefit the public sector (hence centrally paid independent technical assistance is often very welcome).

The public procurement procedure for EPC contracting is a preferred model for street-lighting renovation for local authorities that do not have sufficient funds to finance the renovation in a lump sum, i.e. via the traditional public procurement procedure. Before publishing public procurement tender for street lighting, all local authorities should prepare energy audits of street lighting, which should guide them regarding the technical requirements of street-lighting renovation in line with the regulatory requirements for the area under their supervision. Energy audit needs to be done regardless of the further contracting procedure. In most cases, quantitative criteria contribute between 75 and 85% of the score in the selection of the most favourable bidder. Qualitative criteria vary, but as a rule the municipalities opt for the number of lighting hours, warranty time span, CO_2 reduction, proximity of the service and spare parts availability for a certain period of time as well as maximum response time to change malfunctioning lamps.

Quantitative criteria may encompass only price, i.e. net present value of investment costs, operational costs and financial costs of the private partner or they may include savings in energy and GHG (CO_2) emission expressed in monetary terms in addition. The example of a formula that uses only the price is shown below:

$$NPV_P = \mathrm{CC} + \frac{\sum_{i=1}^{n} (\mathrm{EC} + \mathrm{MC})}{(1+d)^i}$$

Whereby:

NPV_P—the net present value of the overall price of the EPC

CC—upfront capital costs of street-lighting renovation

EC—projected annual electricity costs

MC—projected annual maintenance costs

$i = 1, 2, \ldots, n$ number of years of the EPC

d—discount rate

The example of a typical quantitative criteria might then be:

$0.7\ NPV_P + 0.3\ NPV_S$,

where NPV_S is net present value of energy savings expressed in financial terms. It is more tilted more towards the lower price of the bidder. The balanced choice of bidders would be achieved by attributing 50%

of the weight to each criterion. However, the public authority is free to choose the weight of the selection criteria as long as they respect the quantitative and qualitative criteria rule.

The example of the formula to determine the selection of the bidders according to the quantitative criteria is shown below. It is developed within the EBRD's Regional Energy Efficiency Programme for Western Balkans (http://www.wb-reep.org/). It considers the savings not only during the duration of the contract but the savings after the end of the contract (m) until the end of the expected economic life of renovated street lighting (n)

$$NPV_S = \sum_{i=1}^{m} \frac{(\mathrm{GS} + \mathrm{GM} - N)_i}{(1 + \frac{d}{12})^i} + \sum_{j=m+1}^{n} \frac{\left[(\mathrm{GS} + \mathrm{OU})'\right]_j}{(1 + \frac{d}{12})^j}$$

Whereby:

NPV_S is the net present value of the financial benefit for the local authority, i.e. savings expressed in financial terms

$i = 1, 2, \ldots, m$ is the number of months in which guaranteed savings are realised during the contract term, as proposed by the bidders

$j = m+1, m+2, \ldots, n$ is the number of months from the contract end date until the end of the required minimum economic life of new street lights

$d =$ annual discount rate used for evaluation purposes of the bidders only

GS = guaranteed monthly financial savings from reduced electricity consumption during the EPC contract term, calculated pursuant to the reference electricity consumption and electricity price on the day of publication of the tender as specified in the tender documentation

GM = guaranteed monthly savings on reference maintenance expenditures for the retrofitted street lights, based on the reference electricity consumption, whereby the bidder bears the overall costs of warranty and out-of-warranty maintenance of the new street lights during the contract term, except for the costs of physical replacement of defective lights pursuant to the existing maintenance contract between the local authority and the maintenance company XY for the overall street lighting including those that are not a subject of renovation

GS′ = the expected financial savings arising from reduced monthly electricity consumption in the period after the end of the contract. GS′ shall not exceed GS and shall reflect any reduction in savings from anticipated failures of lights in accordance with their expected operational life

GM′ = expected financial savings arising from reduced monthly maintenance costs in the period after the contract until the end of the economic lifetime. GM′ shall not exceed GM and shall reflect any reduction in savings from anticipated failures of lights in accordance with their expected operational life.

N = regular payments to the bidder in sole consideration for the services provided under the contract. N includes all the costs of ESCO service as requested in the cost list, where $\sum_{i=1}^{m} N = P$, undiscounted price of the services provided by the private partner.

Bidders should disclose all the cost groups, i.e. provide the detailed cost structure, whereby they should respect that monthly fee payable to them by the local authority is in accordance with $N \leq GS + GM$.

These formulas do not consider any financial incentives to the private partner arising from energy savings over the guaranteed level. The intention of most public authorities is to recover the investment cost sooner, rather than later, hence the contracts are typically designed to exploit savings until the investment costs and return to the private partners are reached. The opposite are shared savings contracts in which a part of the savings is used to recover the investment costs and return them to the private partner, while the rest of the savings are used as a surplus to the public budget. The formulas should also be adapted for energy supply contracts, if there are such. If private partners are in charge of energy supply, then it is an additional service that needs to be added to the spectrum of services provided to the public authorities by a private partner.

Governance of Street-Lighting Renovation

The previous discussion mostly dealt with the responsibilities of local public authorities for street lighting functioning and renovation. A common policy of most countries is to set the savings levels in electricity or CO_2 that should be achieved during the certain period of time.

In the absence of national standards, certain EU or international criteria are used (for instance for desirable technical criteria of street lighting). The core governance principle in energy savings is to determine a broad savings goal, yet to allow a bottom-up approach of certain independence of the public authorities in charge of street-lighting renovation. Estimates show that one street lamp is necessary for every nine inhabitants (E-street Project Report, p. 18), which reveals a huge market potential for street-lighting projects and a significant potential for energy savings. Taking into account approximately 750 million inhabitants, the estimated consumption is, based on the same source, calculated as: 80 million light points × 180 W lamp wattage × 4150 burning hours per year = 59,760 TWh per year.

Street-lighting projects are relatively easy to implement, i.e. their implementation can be faster in comparison to other projects. In addition, they are instantly visible to local inhabitants. Hence, it is not surprising that the street-lighting renovation is often conducted before public elections, especially if there are no other impressive projects to demonstrate to citizens.

The greatest savings in electricity consumption and maintenance are achieved when street lamps are replaced according to the 1:1 principle, i.e. one old for one new lamp. In such cases, the private partners have sufficient interest in entering into long-term EPC or PPP contracts if public authorities do not have their own funds to finance lamp replacement. The website of the EBRD's regional energy efficiency programme (www.wb-reep.org) reveals that, for instance, the small Croatian city of Metković achieved as much as 74% electricity and maintenance savings by replacing 1071 old lamps with LED lamps, by entering into an 8-plus year EPC contract with an ESCO. The estimated capital expenditure was about 580,000 EUR, while savings were calculated based on the electricity price of 0087 EUR/kWh. Likewise, savings in the city of Ludbreg have been estimated at 69% due to the replacement of 733 lamps. In both cases, high-pressure sodium/mercury lamps have been replaced by LED technology. The case of the city of Wels in Austria, where the replacement of 9100 lamps was completed between 2011 and 2014, showed a 36% guaranteed annual savings amid the investment costs of 1.66 million EUR and a contract duration of seven years

(Streetlight refurbishment with energy performance contracting 2017). Most estimates show the expected electricity savings between 20 and 50% after LED lamps installation, i.e. in most cases a simple investment payback period is between 2 and 5 years (E-street Project Report, p. 28).

The example of Spain and its municipalities of Girona, showed fewer savings in a wider area with a grant withdrawn from the Cohesion Fund (the website of the so-called Beenergi project is http://beenergi.ddgi.cat/en/resources). While the savings and project payback period depend on the price of the lamps and electricity costs, it is widely accepted that replacement of the lamps, according to the 1:1 principle should not require any grant, as investment costs should be fully recoverable from the savings during the projected economic lifetime of the new lamps (which is typically estimated between 12 and 14 years). However, after the energy market liberalisation, the cost of electric energy in some countries has decreased significantly, compromising the achievement of energy savings in financial terms during the expected lifetime of the LED to recover the investment costs. It also compromises the common stance that no grants are needed for the replacement of one old for one new lamp. For this reason, many larger projects are not financially feasible for the private partners, and grants are needed to bridge the financing gap.

To implement the street-lighting projects and achieve the broad policy goals, municipalities face the choice of technology, timing, scope of the project, budgetary constraints and financing, criteria of the contracting party selection in the public procurement procedure, and local needs. While the regulations were designed to be the same for all, the question is whether all municipalities need the same technical standards of street lighting. It is especially the case for smaller towns whose local authorities face the seasonal or weekend increases of the number of inhabitants. Local budgets vary, and despite the expected savings, there is no universal solution and each local authority should decide what is best for it. By all means, local authorities should strive to achieve value for money, i.e. greater savings, compared to public funds spent for a street-lighting renovation project.

Conclusion

This chapter discussed the importance of street lighting in terms of energy savings, the most commonly used financial models for tackling street-lighting renovation, and some public procurement rules that have been developed to boost energy performance contracting in Europe. Street-lighting renovation is considered to be simple compared to more complex energy savings solutions in public buildings and manufacturing. However, it can be very complex due to a variety of technical criteria, adjoined with the modern lighting systems, rapid changes in technology, varying quality of the luminaries and lamps and smart city solutions. Local authorities are in most cases, incapable of evaluating the technical criteria themselves, without external, independent, technical assistance. Therefore, there is no certainty that public authorities would be able to achieve value for money for the public.

Three financing models have been analysed in this chapter. Traditional public procurement, where only public works are contracted, is the simplest. It is followed by the EPC contracting, while public–private partnership model is, in essence, a no-care model for the public authorities, although it costs more public money.

The financial affordability of local authorities varies. For those that have sufficient funds to cover capital costs, the traditional public procurement model is suggested as the simplest one, while for those that do not have funds at their disposal and cannot borrow them for any reason, energy performance contracting should be a viable alternative to consider.

The public authorities, although bound by national or international policy goals in energy savings, have a choice of determining the size of the investment, the scope of the project, technology, design, timing, private partner selection criteria and financing. All their choices should be made for the benefit of local inhabitants. The flexibility enjoyed by local governments should be monitored by national public authorities, in order to keep the public spending under control. The easiest option is when local authorities change the lamps only, however, it is not always the case. Whenever poles or some installations change, it increases the investment costs. Even though the savings are typically considered to be between 20 and 50% after switching to LED, electricity prices and operation regime, as well as the chosen technical solution, make the comparison of the savings between the municipalities hardly possible. Hence, the ultimate responsibility for achieving best value for money rests with the local public authorities, while local inhabitants should monitor the costs by demanding transparency. Only those projects in which the public knows what they should receive for the money invested can be considered successful.

References

Austrian Energy Agency. (2017). *LED street lighting procurement and design guidelines*. Available at: http://www.premiumlightpro.eu/fileadmin/user_upload/Guidelines/Premium_Light_Pro_Outdoor_LED_Guidelines.pdf. Accessed 17 Feb 2018.

Bertoldi, P., et al. (2014). *ESCO market report 2013*. European Commission, JCR Science and Policy Reports. Available at: http://iet.jrc.ec.europa.eu/energyefficiency/system/tdf/jrc_89550_the_european_esco_market_report_2013_online.pdf?file=1&type=node&id=8869. Accessed 20 Feb 2018.

Boza-Kiss, B., Bertoldi, P., & Economidou, M. (2017). *Energy service companies in the EU*. European Commission, Joint Research Centre. Available at: http://publications.jrc.ec.europa.eu/repository/bitstream/JRC106624/kjna28716enn.pdf. Accessed 25 Feb 2018.

Development Finance International. (2014). *LED road lighting design manual*. Available at: http://www.dfintl.com/wp-content/uploads/2016/04/LED-Road-Lighting-Design-Manual1.pdf. Accessed 21 Feb 2018.

Elberg, R., & Woods, E. (2017). *Smart street lighting as a smart city platform: Applications and connectivity best practices*. Navigant Research, commissioned by Echelon. Available at: https://www.echelon.com/assets/blt339a50e1c88306c2/Navigant%20Research-Echelon%20Smart%20Street%20Lighting%20White%20Paper%20-%20Full%20Report.pdf. Accessed 21 Feb 2018.

Energy Efficiency Directive (EED, 2012/27/EU). Available at: http://eur-lex.europa.eu/legal-content/EN/TXT/PDF/?uri=CELEX:32012L0027&from=EN. Accessed 1 Mar 2018.

E-street Project Report. Available at: https://ec.europa.eu/energy/intelligent/projects/sites/iee-projects/files/projects/documents/e-street_e_street_publishable_report_en.pdf. Accessed 20 June 2018.

European Commission. (2010). *Europe 2020—A European strategy for smart, sustainable and inclusive growth, COM(2010) 2020*. Available at: http://ec.europa.eu/eu2020/pdf/COMPLET%20EN%20BARROSO%20%20%20007%20-%20Europe%202020%20-%20EN%20version.pdf. Accessed 1 Mar 2018.

European Commission. (2013). *Lighting the cities—Accelerating the development of innovative lighting in European cities*. Available at: http://cordis.europa.eu/fp7/ict/photonics/docs/ssl-cip/lighting-the-cities_en.pdf. Accessed 23 Mar 2018.

European Commission (EC)—DG-Environment. (2012). *Street lighting and traffic lights technical background report*. Available at: http://ec.europa.eu/environment/gpp/pdf/tbr/street_lighting_tbr.pdf. Accessed 1 Mar 2018.

Green Partnerships. (2014). *WP3 joint approach for more efficient implementation of local energy strategies, public lighting*. Available at: http://www.greenpartnerships.eu/wp/wp-content/uploads/Green_Partnerships_EWG_PublicLighting2.pdf. Accessed 5 Mar 2018.

Hartley, D., Jurgens, C., & Zatcoff, E. (2009). *Life cycle assessment of streetlight technologies*. University of Pittsburgh. Available at: http://pitt.edu/news2010/Streetlight_Report.pdf. Accessed 10 Mar 2018.

IEA 4E. (2014). *International Energy Agency (Energy efficient end-use equipment) solid state lighting annex task 1—Life cycle assessment of solid state lighting final report*. Available at: http://www.iea-4e.org/document/341/life-cycle-assessment-of-solid-state-lighting-final-report. Accessed 10 Mar 2018.

Matulka, R., & Wood, D. (2013). The history of the light bulb. US Department of Energy. Available at: https://energy.gov/articles/history-light-bulb. Accessed 19 Feb 2018.

Šeba, M. G., & Mušec, M. (2018). Making street lighting ESCO projects work in practice, paper forthcoming in economy, finance and business in Southeastern and Central Europe. In A. Karasavvoglou, S. Goić, P. Polichronidou, & D. Pavlos (Eds.), *Proceedings from the 8th International Conference "The Economics of Balkan and Eastern Europe Countries in the Changed World" (EBEEC 2016)*. Split, Croatia, May 2016.

Street lighting refurbishment with energy performance contracting. (2017). Available at: https://www.plateformesolutionsclimat.org/wp-content/uploads/2015/11/Streetlight-EPC_Guide-European-EN.pdf. Accessed 19 Feb 2018.

The United Nations Environment Programme/Global Environment Facility—UNEP/GEF. (2013). *The rapid transition to energy efficient lighting: An integrated policy approach*. Available at: http://www.enlighten-initiative.org/portals/0/documents/Resources/publications/en.lighten%20Brochure%20PDF.pdf. Accessed 10 Mar 2018.

Traverso, M., Donatello, S., Moons, H., Quintero, R. R., Gama Caldas, M., Wolf, O., et al. (2017). *Revision of the EU green public procurement criteria for street lighting and traffic signals, preliminary report: Final version*. Available at: http://susproc.jrc.ec.europa.eu/Street_lighting_and_Traffic_signs/docs/PR_Final_25-08-2017_Sci4_Pol.pdf. Accessed 20 Mar 2018.

Van Tichelen, P., Geerken, T., Jansen, B., Vanden Bosch, M., Van Hoof, V., Vanhooydonck, L., et al. (2007). *Final report lot 9: Public street lighting, VITO*. Available at: http://www.eup4light.net/assets/pdffiles/Final/VITOEuPStreetLightingFinal.pdf. Accessed 22 Feb 2018.

Velentová, M., Quicheron, M., & Betoldi, P. (2012). *Public procurement of LED lighting in European Member States*. European Commission, Joint Research Centre, Institute for Energy and Transport. Available at: http://publications.jrc.ec.europa.eu/repository/bitstream/JRC69696/public_procurement_of_ssl_new_layout_v2online%200.pdf. Accessed 8 Mar 2018.

Welz, T., Hischier, R., & Hilty, L. M. (2011). Environmental impacts of lighting technologies—Life cycle assessment and sensitivity analysis. *Environmental Impact Assessment Review, 31*, 334–343.

Part III

Conclusion

10

Renewable Energy and Resource Efficiency: Governance Is Key

Nikolai Mouraviev and Anastasia Koulouri

This concluding chapter looks at the big picture: it draws insights into the nature of energy security; identifies critical questions requiring further investigation and debate; highlights the value of the energy security model offered in this book; and discusses broader considerations and future research opportunities.

The Nature of Energy Security: Contextual, Dynamic, Multidimensional and Polysemic

Considering the inherent nature of energy security, as exhibited by its conceptualisation since its emergence, it is worth emphasising its *dynamic* character that evolves over time as socio-economic,

N. Mouraviev (✉) · A. Koulouri
Dundee Business School, Abertay University, Dundee, UK
e-mail: n.mouraviev@abertay.ac.uk

A. Koulouri
e-mail: a.koulouri@abertay.ac.uk

N. Mouraviev and A. Koulouri (eds.), *Energy Security*,
https://doi.org/10.1007/978-3-030-01033-1_10

geopolitical and military conditions and priorities change, our knowledge and understanding of the natural world grow, and technology advances (Ang et al. 2015). The meaning of energy security is also *contextual* as its understanding and conceptualisation are heavily dependent on the prevailing socio-economic and political paradigms and existing conditions; local, regional, national and international circumstances; and a range of risks, their levels and potential impact (Ang et al. 2015; Dannreuther 2017). Furthermore, energy security is *multidimensional* touching upon and/or being influenced by various issues, such as resource availability and accessibility; technological capacity; policy direction, design and development; policy implementation; economic sustainability; and environmental impact (Baumann 2008; Chester 2010; Vivoda 2010). Finally, energy security is *polysemic* (Chester 2010), having multiple meanings. For example, its meaning is different for energy importers vs. energy exporters, OECD states vs. emerging economies, industrialised countries vs. low-income nations. Therefore, there is no universally accepted definition of energy security, but rather a continuously evolving concept drawing variable levels of attention and intensity of activity by academics, practitioners and policy makers (Ang et al. 2015; Kuzemko et al. 2016).

Critical Issues

The investigation of the energy security concept, its underpinning assumptions, its dimensions and the government agenda seeking to enhance it reveals the evolution of the concept, as well as development, often controversial and/or inconsistent, of the energy policy and practice in nations across the globe. It also highlights a range of issues that continue to draw the attention of academics and practitioners and that need to be addressed by governments and other actors in the energy sector and beyond. The most salient questions that still require answering include:

- What are the criteria that governments use for (in)forming their views on energy security? Often energy security is driven by the political forces and their agendas, although the principal concerns do not relate to politics, but rather to economic and environmental sustainability.

- Although the national context plays a significant role in shaping energy policy and the meaning of security, it also often creates path dependency that many nations are struggling to break. More generally, looking beyond the context, is there something more influential (i.e. more significant than the context) that forms the core of energy security? If so, the context would appear as a limiting factor that often comes to the fore but does not allow a nation to adopt a long-term perspective on energy and energy security.
- Can governments learn from their own experience to date? Again, not only path dependency sets constraints on what the governments could do to enhance energy security, but also limits the learning opportunities by steering an agenda in the direction of existing, continuous problems typically related to politics and the supply and use of fossil fuels.
- What are the benefits, drawbacks and/or trade-offs involved in using what this book called the emergent twenty-first century energy security paradigm (see Chapter 2), that implies a multidimensional approach? Is there a threat to energy security that comes from the paradigm itself? To elaborate, should the emergent multidimensional approach to energy security be viewed as inherently beneficial due to its holistic nature and its attempt to assemble many items of varying importance, related to energy security? Or, on the contrary, is its value fundamentally limited as its focus is diluted? Embracing many, perhaps too many, elements trying to capture every single detail related to energy security presents an enormous challenge for governance. What should governance address first, how are the tasks prioritised, will items at the bottom of the list ever receive government attention, are there enough financial resources and administrative capacity to tackle energy security in its multidimensional meaning? Critically, does the multidimensional approach increase or decrease the government's ability to enhance energy security?

These and many other issues regarding energy security require further investigation and are subject for discussion as many aspects would still benefit from deeper conceptualisation. Currently, there is a large variety of perspectives, complementary and/or competing, and energy security should be viewed as a rapidly developing research field.

The Energy Security Model

This book offered its own insights into some critical areas and contributed to an ongoing debate. The book's view on energy security focuses on two crucial parts: renewable energy and resource use efficiency. We view these two parts in their dynamic interaction: one drives the other, and both should be approached as long-term tasks with progressively increasing and, likely, continuously challenging targets. Energy generation from renewables should be gradually increasing, thus replacing energy from fossil fuels, whilst continuous benefits to the energy system and society at large should also be achieved by setting gradually increasing targets for improved utilisation of energy resources.

We adopted the governance perspective to make our conceptualisation of energy security more practical: the rhetoric regarding renewables and efficiency is not going to help unless government policy objectives are supported by an elaborate set of governance structures, investment schemes, incentives, subsidies, attractive tariffs, well-designed procedures and effective processes for energy producers and consumers.

Anyone who studies energy security is aware that this concept is also a buzzword that catches the attention of many. Using security jargon is helpful to draw the initial attention of international organisations, policy makers and the general public. However, it is not enough for strengthening energy security. The book's governance perspective on energy security permits—and might be very useful—when considering the risks associated with international relations and seeks not only political, but also managerial solutions that would mitigate existing or perceived threats arising from international tensions. By enhancing RES utilisation and resource efficiency via consistent governance, a nation becomes less vulnerable to domestic, as well as international, risks. What might be viewed as threatening in relation to renewables and energy efficiency is often continuous, a persistent lack of attention paid to making governance effective.

Does the proposed approach disregard or diminish the role of the context? Whilst renewables and resource efficiency might remain common goals for many nations, both resource-poor and resource-rich, the details of governance and its targets will be highly contextual.

It is likely that resource-rich nations will adopt a slow (or slower) approach to ever-increasing energy generation from renewables and might be not very active in pursuing energy efficiency programmes (the experience of Kazakhstan, Russia and Ukraine highlighted in the book shows this). What is important is the policy direction, supported by proper governance, although one might realise that implementation might take decades. As energy security has both national and international dimensions, coordinated governance is required both domestically and internationally (Szulecki 2018) with likely considerable variation in governance tools used by nations across the globe.

The book did not look for the 'true meaning' or the single best definition of energy security. Rather, we adopted the long-term perspective focused on sustainability, and this theoretical framework allowed us to create a model that rests on two critical tasks for the energy sector and focuses on the gradual reduction of dependency on fossil fuels. Ultimately, the model's value is in its ability to transform the energy sector and ensure sustainability. The sustainability approach is useful as it accommodates critical perspectives on energy security (renewables and energy efficiency), which forms the governance' focus. However, we acknowledge that there might be other models, based on different theories or concepts, that would also enable society's sustainability.

Broader Considerations and Future Research

It is likely that internal drivers, such as path dependency, powerful oil sector lobby, vested interest in existing contracts and historically formed supply chains, will continue to exert pressure on governments for a lengthy period of time, shaping their views and actions regarding energy security. This brings up a question about values related to energy security (Cherp and Jewell 2014). Does energy security need to promote certain values and/or conform to some priorities and interests and/or strengthen certain values? If so, which ones? Who sets these values and how does society know that these values are beneficial? Although the book does not deal with values directly, it nonetheless provides a general answer that departs from narrowly defined short-term political and economic interests. The principal value that could be achieved by the ever-increasing utilisation of renewable energy, complemented by continuous improvements in efficiency, is sustainability. Whilst sustainable development delivers economic, social and environmental gains, a detailed discussion of how these gains manifest themselves and how they could be enhanced is beyond this book's scope.

Various kinds of benefits stemming from strengthened energy security present a fruitful opportunity and a theme for future research: energy security as value to society. In addition to the critical issues noted in the beginning of this chapter, other topics for investigation include assessment of the trade-offs facing the government in selecting governance instruments; partner interaction in various governance settings; conditions for and the direction of a paradigm shift in conceptualising energy security; and the role of civil society in strengthening energy security. The active engagement of citizens, interest groups, community associations and entrepreneurs (e.g. methods of local community mobilisation for off-grid and grid-connected power generation from renewables) is yet another opportunity not to be missed in both research and governance.

References

Ang, B. W., Choong, W. L., & Ng, T. S. (2015). Energy security: Definitions, dimensions and indexes. *Renewable and Sustainable Energy Reviews, 42,* 1077–1093.

Baumann, F. (2008). *Energy security as multidimensional concept.* Research Group on European Affairs, No. 1. March 2008. Centre for Applied Policy Research (CAP), Munich. http://www.cap.lmu.de/download/2008/CAP-Policy-Analysis-2008-01.pdf. Accessed 7 July 2018.

Cherp, A., & Jewell, J. (2014). The concept of energy security: Beyond the four As. *Energy Policy, 75,* 415–421.

Chester, L. (2010). Conceptualising energy security and making explicit its polysemic nature. *Energy Policy, 38*(2), 887–895.

Dannreuther, R. (2017). *Energy security.* Cambridge: Polity Press.

Kuzemko, C., Keating, M. F., & Goldthau, A. (2016). *The global energy challenge: Environment, development and security*. London: Palgrave.
Szulecki, K. (Ed.). (2018). *Energy security in Europe: Divergent perceptions and policy challenges*. London: Palgrave Macmillan.
Vivoda, V. (2010). Evaluating energy security in the Asia-Pacific region: A novel methodological approach. *Energy Policy, 38*(9), 5258–5263.

Index

N. Mouraviev and A. Koulouri (eds.), *Energy Security*,
https://doi.org/10.1007/978-3-030-01033-1

FSC
www.fsc.org
MIX
Papier | Fördert
gute Waldnutzung
FSC® C083411

Kolibri 360 GmbH
Ferdinand-Jühlke-Straße 7
99095 Erfurt, Deutschland
produktsicherheit@kolibri360.de